Students' Laboratory Manual of Physical Geography

Albert Perry Brigham

Alpha Editions

This edition published in 2020

ISBN : 9789354006142

Design and Setting By
Alpha Editions
email - alphaedis@gmail.com

TWENTIETH CENTURY TEXT-BOOKS

STUDENTS' LABORATORY MANUAL

OF

PHYSICAL GEOGRAPHY

BY

ALBERT PERRY BRIGHAM

PROFESSOR OF GEOLOGY IN COLGATE UNIVERSITY

NEW YORK
D. APPLETON AND COMPANY
1905

PREFACE

THE exercises in this manual follow the order of Gilbert and Brigham's Introduction to Physical Geography, bearing, as in the Teacher's Guide, corresponding section numbers. The figures mentioned belong to the Introduction, unless specified as found in the Manual. A few views and diagrams are here introduced as a basis for laboratory work. They supplement those of the text and may serve to test the value of such material for this use.

About one hundred and thirty exercises are given, but it is not expected that any class will accomplish all of the work proposed. It is left for teachers to select according to the material and time available. Other exercises might in many cases have been prepared, but the choice has been given to subjects whose illustration by maps or other means would require small expense. After a few added years of experience in the schools, we may hope to approach what might be called a standard series of exercises, though locality will always be a large factor, and will determine the subjects for work in the field. It is obvious also that the directions given for field studies must be of a general nature. It is suggested that the index-maps showing progress of topographic work in the several States would be useful in enabling the student to plot the area of sheets studied, upon the general map of the United States. They can be obtained from the United States Geological Survey, either separately or included in the annual reports.

The brief treatment of life in this manual in no way represents the author's estimate of its importance. The individual teacher must lead in supplementing the text on this theme, since the material is largely in books, and the selection depends on the school library. A few exercises are given, based on maps which show distribution.

ALBERT PERRY BRIGHAM.

COLGATE UNIVERSITY, *September*, 1904.

CONTENTS

LIST OF ILLUSTRATIONS

A LABORATORY MANUAL OF PHYSICAL GEOGRAPHY

CHAPTER I

THE EARTH

1. **Curvature of the Earth's Surface.*** — Magellan's nationality, place and date of birth? Any facts about his history previous to his great voyage? Under what king was his expedition organized? How many ships had he? Date of sailing? Trace the earlier course of his voyage. What islands in the Atlantic did he touch? What point on the African shore did he make? Where did he first touch the South American coast? Distance between the two continents at these two points? What was his uncertainty at the mouth of the La Plata? At what port did he winter? What serious trouble happened here? When did they reach the Strait of Magellan? How long were they in passing through it? What did Magellan call the great sea on which he entered? In what direction did they at first sail after issuing from the strait? What sufferings

* Atlas. Encyclopedias. Outline Map of the World. Discovery of America, John Fiske, vol. ii, 184–212.

If Fiske is not at hand use the following list of places touched in Magellan's voyage: Canaries, Sierra Leone, Pernambuco, La Plata, Port St. Julian, Cape Virgins, San Pablo, Ladrone Islands, Philippine Islands, Borneo, Moluccas, Cape of Good Hope. The Encyclopedias, Art. Magellan, give the more important events to which reference is made.

did they undergo in crossing the Pacific? How were these relieved at the Ladrones? What befell the commander and some of his captains upon one of the Philippine Islands? How near had Magellan approached this point in a former voyage from the west? How many ships reached the Cape of Good Hope? How many survivors reached Spain? How long a time did the circumnavigation of the globe occupy? How long would it require to-day? If an outline map of the world is at hand, plot the course of Magellan's entire voyage upon it.

Have you observed an eclipse of the moon? What was the shape of the shadow cast by the earth on the moon's face? Could it have been made by other than a round body? Answer this question by considering the shadows made on the wall by various bodies put between the wall and a lamp. Try a brick, a cobblestone, an egg, a dinner-plate, a tin pail, a ball. How many of these would cast a circular shadow in some position? How many in all positions? Would the observation of one eclipse prove the spherical form of the earth? Would the observation of many such eclipses make the conclusions quite certain? How far did Magellan's voyage go toward proving the earth a globe?

In Fig. 1 of the text-book how do the ships on the horizon differ from those near at hand? Would this be true anywhere on the sea, of ships seen at different distances? Would the same be true on large lakes? Suppose you stood on the deck of a ship in mid-ocean, on a clear day when the water was smooth, would the horizon seem equi-distant in all directions? Would this help us judge as to the form of the sea surface? How would the horizon appear if the earth had the shape of a cylinder?

If one stood at the Equator he would see the Pole Star on the northern horizon. If he stood at the North Pole he would see it in the zenith. Suppose the observer, going at a uniform rate from the Equator to the Pole, should see the height of the Pole Star increase at a practically uniform rate also. Could he draw a conclusion as to the form of the earth? Consider an egg-shaped earth in the same way. Would the angular height of the Pole Star increase uniformly to one traveling from the Equator to the Pole? Where would the star increase in altitude very slowly? Where quite rapidly?

If you had never before learned that the earth was round, do you think any one of the above arguments would convince you? What seems to you the strongest proof? How would the force of all these considerations taken together affect you as compared with any one taken singly?

2. **Study of Latitude and Longitude by the Globe or Map.**—What cities of over 100,000 inhabitants are found in the United States within 2° of 40° north latitude? Are any important cities of Europe near 40°? What countries are crossed by this line? What important cities of Europe lie within 2° of 50° north latitude? Has North America any important center of population as far north as 50° north latitude? What cities has Europe near the sixtieth parallel? What is the latitude of Cape Farewell? How does Asia compare with Europe and North America as regards the grouping of great cities in a single belt of latitude? Has the world any great cities within 10° of the Equator? What city of South America is at an equal distance from the Equator with Winnipeg? How do Melbourne and Louisville compare in latitude? What is the difference of longitude between the two cities? Find on a globe the point antipodal to yourself. How is the point antipodal to New York located with reference to Melbourne and Cape Town? Recalling the experience of Magellan in the strait bearing his name, compare its latitude with that of Edinburgh. How strongly does latitude determine climate? (It is not expected that the student at

this point will give more than a partial answer to this question.) Compare the longitude of Sitka and of Behring Strait with that of the Isthmus of Panama.

*2a.** **Variation in the Length of a Degree of Latitude or Longitude.**—If the globe were a perfect sphere, a degree of latitude would be the same near either pole as near the Equator. But, owing to the polar flattening, a degree is the arc of a larger circle in the far north and south, and hence is a little longer. In section 2 the length of a degree of latitude is given as *about* 69 miles. The real variation is less than a mile, or from about 68.7 miles at the Equator to 69.4 miles at the poles.

The student will now consider the degree of longitude. Suppose two meridians a degree apart. They cut the Equator 69.172 miles apart; but they come together at the poles. The angle between them is everywhere the same, 1°, but the included arc is 1-360 of a smaller and smaller circle as we approach either pole. Below is given the length of a degree of longitude for each successive tenth degree of latitude:

LATITUDE.	DEGREE OF LONGITUDE.	LATITUDE.	DEGREE OF LONGITUDE.
0°	69.172 miles	50°	44.552 miles
10°	68.129 "	60°	34.674 "
20°	65.026 "	70°	23.729 "
30°	59.956 "	80°	12.051 "
40°	53.063 "	90°	00.000 "

Using the above data, answer the following questions:

How far is Denver, Col., from Philadelphia? How far is it from St. Augustine, Fla., to New Orleans? What

* The letters *a*, *b*, etc., do not refer to corresponding subdivisions in the Introduction, but designate alternative or supplementary exercises given in connection with many sections of the text-book.

is the distance from Christiania to St. Petersburg? How much difference in the longitude of the two capitals? Distance from Prag to Land's End? Difference in longitude? Distance from Philadelphia to the point where the Nebraska-Kansas boundary touches the Missouri River? Difference in longitude? What is the difference between Madrid and Constantinople? (They may be considered as

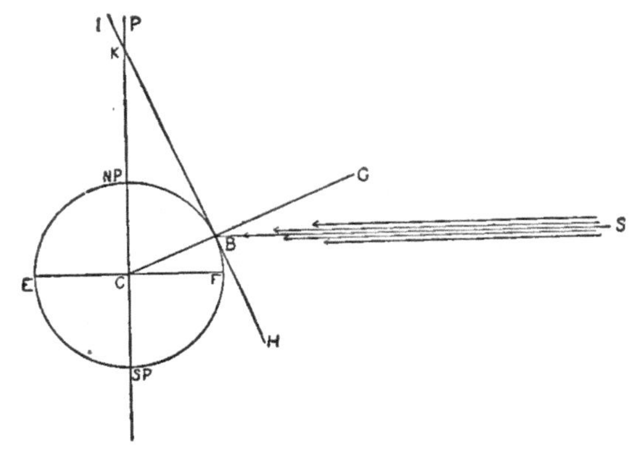

FIG. 1.—Diagram to show that latitude = angular distance of Pole star above horizon; it also represents the relation of the earth to the sun at either equinox.

of latitude 40° north.) How far apart do their meridians cut the Equator?

2c. **Determination of Latitude.**—Chapter II of the text should be studied before this exercise is taken, unless the student is familiar with the principles there given. In Fig. 1 of this manual

EF represents the Equator;
P represents the North Pole of the Heavens;
B represents the point whose latitude is to be found;
IH represents the Horizon of the observer at that point;
G represents the Zenith.

The arc BF or the angle BCF represents the latitude.

By geometry, angle BCF = angle CKB = angle IKP. Now this last angle IKP is the angular distance of the Pole of the Heavens above the observer's horizon. Let the student construct other figures like this, letting point B be anywhere between F and NP, and thus see that this conclusion always holds true—viz., that the latitude of any place = the angular distance between the horizon and the North Pole of the Heavens. The student is now prepared to understand the *principle* on which the mariner and explorer may determine latitude. On the sea, since its surface gives a perfect horizon, the task would be easy if the North Star were always in sight. But the sailor is not shut up to this. He knows by the tables of the nautical almanac how far the sun and many stars are from the North Pole of the Heavens, and he then computes the angle IKP, and this, as we have seen, equals the latitude.

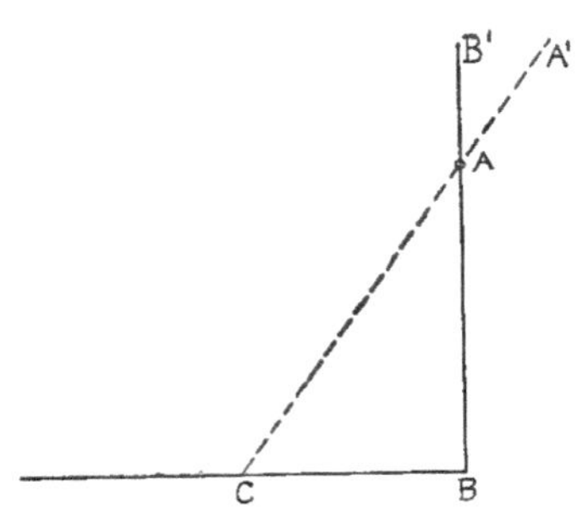

Fig. 2.—Diagram to illustrate a method of finding latitude.

A rough measurement of the latitude may be made in the schoolroom, if it has a south window, at the time of either equinox (March 21st or September 23d). The sun at noon is then 90° from the Pole of the Heavens, and the angle between the sun's direction and the vertical line equals the latitude. This also will appear in Fig. 1 if we now regard it as representing the earth at an equinox. The sun's rays will reach B from S, and the angle between the sun and the zenith, SBG, equals, by geometry, angle FCB, which is the latitude.

Hang a plumb-line close to the window; mark the point

directly below it on the floor (B), Fig. 2 of this manual; call that the base-point; attach a button $\frac{1}{2}$ inch wide at A, 100 (measured) inches above the floor (if the window is too low, substitute 50 inches, and make allowance in the computation). Draw a line on the floor northward from the base-point. If the direction of true north is not known, find it by means of a compass and the declination, as shown by Fig. 189, or else find it (on a previous day) by the principle that the shadow of the button is nearest the base-point when directly north from it. On the day of the equinox, when the shadow is on the north line, mark its position (C) and measure in inches the distance to the base-point.

We have now a right-angled triangle, ABC, in which we know the sides AB and BC. By trigonometry, therefore, we can find the size of the angle BAC. This angle is equal to angle B′AA′, which as we have seen above = latitude. The trigonometrical formula is, tangent of latitude $= \frac{CB}{100}$, but as the trigonometrical tables may not be convenient, we give a table showing the values of CB corresponding to each fifth degree of latitude. Thus if CB is more than 70 inches, but less than 83.9 inches, the latitude is between 35° and 40°.

LATITUDE.	CB.	LATITUDE	CB.
0°	0.0 inches	50°	119.2 inches
5°	8.7 "	55°	142.8 "
10°	17.6 "	60°	173.2 "
15°	26.8 "	65°	214.5 "
20°	36.4 "	70°	274.7 "
25°	46.6 "	75°	373.2 "
30°	57.7 "	80°	567.1 "
35°	70.0 "	85°	1,143.0 "
40°	83.9 "	90°	Infinite
45°	100.0 "		

13. **Contoured Maps.**—Mount Shasta Sheet, California; or Topographical Atlas of the United States, Folio I. Why is this called the Mount Shasta sheet? Observe that the land area which such a map represents is commonly called a quadrangle. Be careful therefore to use quadrangle for the land and sheet for the map. Observe the figures at each corner. Which show latitude? Which longitude? By means of them locate the position of this quadrangle on a map of the United States or of the Pacific coast. To what range of mountains or series of peaks does Mount Shasta belong? See reference in folio. What is the scale of this map? See the margin at the bottom. What distance stands for 1 mile? How long is the map, compared with the length of the quadrangle? See fraction under head of scale at the bottom. Does this map show much or little country, as compared with most maps that you have seen? Is it a large- or small-scale map? Compare scales in other maps, as in a school geography. Is there advantage in a large-scale map? How many square miles are included in this quadrangle?

What does this map show? To find this out we must learn to "read" the map. Observe the word "Legend" on the left margin. This explains the signs that are used. Could different signs have been chosen? Observe that the signs are given in three groups—under Relief, Drainage, and Culture. Let us take the culture signs. How many are there? What do they show? Would any of these features have been in the quadrangle if man had not been there? In what color are these features shown? Is anything else on the map printed in the same color? If so, what? Are there any towns on the quadrangle? What sign is used for highways? Are these confined to any part of the district? There is but one railway here. Has it an unusual course? Is there any other kind of way for the passage of man or beast? What other culture feature is indicated? Where is it on the map? Why at this point? Note that surveyors in a new country select high points first for the use of their instruments. Do many people live in this quadrangle? How do you know? Make an estimate of the number from the number of the houses.

We will observe next the drainage. In what color are the signs printed? How many signs are there? Are any large streams here? Measure the length (on this map) of the longest stream you can find. On what plan are the streams arranged? Compare the northwest and southeast sides of the mountain and write down your observation. Can you tell whether any of the streams rise close to the top of the mountain? What do the dotted blue lines show? What are such streams called in the legend? Have you seen stream-beds that only carried water in certain seasons or after storms? Notice the sign for "sink." This is a place where a surface stream disappears and goes down into the earth or rocks below. Locate such a sink on the map. Are any large lakes here? Are there ponds? Are they of different kinds? Locate one or more of each. Is there any other drainage feature? What is its name? How many of this class are named? Where are they on the slopes of the mountain? A glacier is a mass of ice made of packed snows and creeping slowly forward. Such ice even on a high mountain slowly melts. What becomes of the water? Do you see any correspondence between the plan of the streams and the position of the glaciers on Mount Shasta?

The Relief.—By this we mean that some parts are higher than others. Points of different height are connected by slopes, steep or gentle, or sometimes by vertical cliffs. Relief is shown in these maps by signs known as contours. Observe the brown lines on the map of Mount Shasta. In a general way they run around the point which marks the summit. Are any of the brown lines heavier than the others? How often do the heavy lines replace the light ones? Do you see any breaks in the heavy contours? What is in these gaps? Find the heavy contour which shows the number 7,000. Begin at any point and trace this line around the mountain until you come back to the point of beginning. The line represents a level tracing around the mountain, which is everywhere 7,000 feet above the level of the sea. We call it the 7,000-foot contour. Suppose a real road were built around the mountain at this level, and you had a bird's-eye view of it from above the summit. This is the appearance given by the contour on the map.

What is the contour interval? See margin at bottom. Moving in toward the center (summit of the mountain), what altitude does the next heavy line represent, counting each interval as 100 feet? Is this line numbered? What altitude does the next heavy contour represent? Is it numbered? Is there a need of numbering all the lines? What is the advantage of making every fifth contour heavy? Moving from the 7,000-foot contour outward, or down the mountain, what altitude for the next heavy contour, and the next? What is the lowest point that you can determine on the map? What is the highest? What is the difference between the two? This is the total "relief" of this quadrangle.

Go back to the 7,000-foot contour. It is curving, almost everywhere. Would this be true if you were walking around a hill? If it were a smooth hill standing by itself, the line would everywhere curve outward. If there were small, shallow valleys running down the hill, would the level line bend inward? If there were a ravine running down the slope, what kind of a turn would a level line make that should run back into it? Do you notice any sharp turns inward in tracing this contour? Do the lines above it turn in the same way? The lines below it?

Observe the contours along Mud Creek. How do they cross the line of the stream below 4,000 feet? Do they begin to turn across it in a different way above 4,500 feet? Is the angle sharp or blunt? How is it above 6,000 feet? How is it above 9,500 feet? How do the contours turn just below the Konwakiton Glacier? Do these differences teach us anything about the character of the valley at the different points as it descends the slope of the mountain? Taking the contours on this map as a whole, what do the outward curves represent? What do the inward curves and turns represent? Observe that above Horse Camp there are sharp turns pointing outward on either side the Sisson trail. This means that sharp ridges or spurs stand out from the summit at this point. Thus a contour runs in or out, by a curve or an angle, according to the form of the mountain.

Notice the number 14,380 at the summit. This marks its altitude. Observe Shastina to the westward, and 12,433 marking its altitude. Follow the trail from the triangulation station westward until you come to the 12,000-foot line. Go a little farther west until you come to another 12,000-foot line. Still follow the trail westward, and it brings you up near the summit of Shastina to a small crater lake. The space between the two summits may be called a "saddle." Would the lowest point in the saddle be much below 12,000 feet? Study Gray Butte. What is its altitude? How much would you have to go down in order to ascend to the summit of Mount Shasta from the top of Gray Butte? Study the cinder cones and Bear Butte in the same way. What does the smoothness of the lines representing these cones show about their form?

On what part of the map are the contours close together? What does this mean? Where are they farther apart? What does this mean? What river crosses the quadrangle on the southeast? Which way is it flowing? What is the altitude of the Southern Pacific Railway where it enters the quadrangle on the south? What is its altitude at the bend a little to the north? How far apart are these two points? How far is the line carried to make this distance?

Notice in the legend the symbol used to represent cliffs. These short parallel lines are often used to represent mountains and slopes. Why are contour lines more useful? These short lines are called hachures, and are used here with the contours, because the cliffs are so steep and so high that many contours would be needed at the same point. Thus all would run in together and be confused. Where are such cliffs found on Mount Shasta?

To draw a profile of the mountain across the summit. Take a strip of white paper at least 17 inches long and 4 inches wide. Place it so that the edge stretches from the northwest corner of the map through the point 20′ on the east side. It will also pass across the triangulation station at the summit. Mark the edge of the paper at every point where it is intersected by a contour line of even thousands of feet. What is the lowest such line on the northwest? On the southeast? Also mark the edge where it passes off the edge of the map (at the corner and at 20′). Let 1 inch to the mile be the vertical as well as the horizontal scale; $\frac{1}{5}$ of an inch will then stand for about 1,000 feet. Where the 4,000-foot line touches the edge of the paper, raise a vertical line $\frac{4}{5}$ of an inch in height. At the crossing of the 5,000-foot line make the vertical 1 inch, and so on. At the summit the mountain lacks about 1,000 feet of being 3 miles above the sea. Make the vertical nearly 3 inches. Go down the southeast side in the same way. Raise verticals at the edge, a little less than $\frac{4}{5}$ of an inch at the northwest, and a little less than 1 inch at the southeast. Connect summits of all the verticals with a gently curved line. This will be a profile of the mountain in this direction. Would any other direction give a similar result? What level is represented by the bottom or lower edge of the paper? Compare this profile with that seen in the view of Mount Shasta (Fig. 144).

Mount Shasta is called a "young volcanic mountain." It has been built up gradually of molten and crushed rock issuing from the crater or craters of a volcano. This work has ceased, but the streams of water have not cut many gorges down the side of the mountain, though a few are deep. How deep is the gorge of Mud Creek where the 6,000-foot line crosses the stream? Some flows of lava still show where they came to rest. They have not been much worn or wasted since. These facts teach us that the glaciers and frosts and streams have not been at work cutting the face of the mountain for more than a short period. It may therefore be called young. See exercise 185.

13*a*. **Fargo Sheet, Minnesota, North Dakota.**—Why is this called the Fargo sheet? Where is this area located—latitude, longitude, State? Note the figures at each corner, and by means of them mark the location of this region on an outline map of the United States, or of the States in which it belongs. What is the scale of the map? See margin below. What distance on the map represents a mile in the region? How many feet in a mile? In a kilometer? Why two scales? What proportion does the height or width of the map bear to the extent of the quadrangle from north to south, or from east to west? See fraction below map. Is this, for a map, a large or small scale? Compare any map in your school atlas. Is there any advantage in a large-scale map? How many square miles are included in this area?

What the map shows. We will notice first the "culture" features. The introduction has told in a general way what these are—the more especial work of man. How many towns of some size, and where? What signs are used for roads? What are the common directions of the roads? How far apart as a rule? What symbol for dwellings? Why no such symbols in Fargo? Draw in your notes a sample of the county boundary-line (see legend at right side of map). How many counties are shown in part on the map? How many States? See legend for State boundary. As here it follows the crooked course of the Red River, the line appears blurred and indistinct. Draw, in your notes, a sample of the township boundary. If a road follows a town line or county line, how is it shown? What are the "section" lines, and what do they mean? Why is the name "Casselton" printed on the west border of the map? See last items of legend. No such name appears north, east, or south, because the adjoining maps on those sides were not yet completed. Symbol for railways, and names of railways in this area? What features of the map are printed in black?

We will examine the drainage. Are there many streams or few? In what general direction do they flow? How can you tell? What is the chief stream? Its principal branches? Do they all join the chief river within the area here shown? What is meant by the dotted blue lines? See legend under drainage. How does the symbol of the large streams differ from that of the small ones? Are there any lakes? If so, how large are they? Are there any depressions or basins that are without water? See legend and then search the map.

Study of the relief of the Fargo Quadrangle. See the explanation of contours given in section 13 of this manual and the account of the Topographic Map in Folio 1. What is the contour interval upon this map? What does this mean? In what color are the contour lines printed? What advantage in a color different from those of the drainage and culture features? Note the words at the bottom of the map defining the "Datum." We mean by this the level or plane which is given, or from which we begin and work up. What do we mean by the 900-foot contour? Find it at any point on the map. Trace it throughout its course, and describe it in your notes, with reference to streams, towns, or other features. Why does it run out to the westward in a double-headed spur about 3 miles south of Fargo? Notice any other points where it bends in like manner. From the direction of the streams where would you look for contours below 900? Where is the 880-foot contour? Where the 860-foot contour? Why so close together? Is there any lower contour on the map? Why does the 900-foot contour swing off to the west and east northward from Fargo. How far north of Fargo is the surface of the Red River exactly 860 feet above sea level? Where is probably the lowest point in the quadrangle? Can we tell how much below 860 feet it lies? Is it as low as 840 feet? The 900-foot line is still close to the Red River at the south end of the map. How much does the river descend in this quadrangle? Is this section of the river much longer than the quadrangle? How do you know this? If the river is twice as long in its actual course as the quadrangle, what is its fall per mile? What are these curves in a river called? Would they be found along a stream that descends rapidly? Are there any points along the Red River where "cut-offs" seem likely to take place—that is, where the parts of a river on two sides of a bend come so close together that the stream may cut through or cut off the land within the bend? How much is the river sunk below the general surface—toward the north—toward the south—and for a few miles north of Hickson? Is this channel wide or narrow?

We have seen that the streams flow sluggishly to the northward. How far apart are the contour lines back from the streams? Can you find places where areas more than

3 inches wide are not crossed by contours? How many miles could you travel east or west without rising or going down more than 20 feet? How many miles northward or southward? Does the 1,000-foot line appear on the map, and if so, where? What is the highest line, and where? Is there any hilly region in the quadrangle, and if so, where? Could you tell by the eye, if traveling here, that most of this quadrangle was not absolutely level? Are there any roads on diagonal lines in this area? Would much attention need to be paid to grades in laying out a railway line here?

Would some other location have been as good for Fargo and Moorhead? If so, where? Are the country homes scattered evenly or unevenly over the area? Would this be likely to be so in a region of hills and valleys? Could water-power be used here? Would the conditions be favorable for mills of any kind? The references will show that this is a region of very rich soil. We have seen that it is flat and easily cultivated. What is the leading industry sure to be? The surrounding lands, especially far up and down the Red River, are like these. The railways and highways can be laid out anywhere at convenience. Town sites are likely to be determined by convenient points on a railway. Here such a point is by a river, and the occasion is partly natural and partly due to man.

We have seen that the area is flat and the streams are few and sluggish and the soil is mellow and deep. If there were steep slopes, or thin soil, would there be more streams, or larger streams, and would they cut deeper valleys? If these streams can work long enough, will they dig out deeper and wider valleys? If streams can do much work, and have done but little, they must be young streams, which have not been flowing for a long period where they now are. This we know to be true in the Fargo region, and we call it young or immature.

Can you find any features on this map to which your attention has not been called? If so, note them.

Comparison of Mount Shasta and Fargo Sheets. Mark the location of the two sheets on an outline map of the United States. How does the scale of the Fargo Sheet compare with that of the Mount Shasta Sheet? With this difference in scale, how much more land in one quadrangle than in the other? How many signs are used for "Culture" or human features in one? How many in the other? Is the same color used in both cases? Why? Are there any culture signs on the Shasta Sheet that are not on the Fargo Sheet? Why? Can you compare the population of the two quadrangles?

What drainage features of Shasta are not found in the Fargo region? How do the streams of the two quadrangles differ from each other, in direction, and in directness of flow? What is the difference in contour intervals of the two maps? Why is the larger interval suitable for Shasta? Why is the smaller interval suitable for the Fargo map? Would 100-foot contours tell you much about the surface around Fargo?

CHAPTER II

THE EARTH AND THE SUN

18. **The Earth's Orbit.**—It will be best to begin by drawing an ellipse whose major axis much exceeds its minor, and come gradually to one representing the form of the earth's path.

Draw an ellipse 8 inches long and 5 inches wide. To

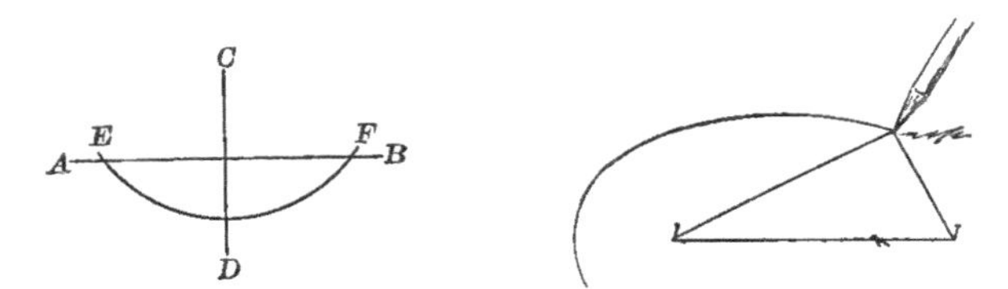

Fig. 3.—Diagrams to illustrate the drawing of an ellipse.

do this, lay off two lines, AB and CD, Fig. 3 of this manual, representing the length and width, so that they cross at right angles and at the middle of each. With C as a center and with a radius of 4 inches (half of AB) draw an arc intersecting AB in E and F. These points are to be the foci of the ellipse. Set a pin at each. Make a loop of string just long enough so that when passed about the two pins it can be drawn out to C or A. Place a pencil inside, and pressing outward to keep the string taut, draw with it a curve. The curve is the desired ellipse. Draw ellipses of other forms.

Draw an ellipse of the form of the earth's orbit. For this the length of the string should be sixty-one times as great as the distance between the foci. Draw a circle about one of the foci, and observe how little the two curves differ.

21. **The Seasons.**—On a broad table (or on a broad sheet of paper lying on a table) represent the earth's orbit by a circle with a radius of 3 feet. A marble $\frac{1}{3}$ of an inch in diameter may be placed at the center to represent the sun on the same scale. An ordinary grain of table salt is too large to represent the earth.

Make 12 equi-distant points on the circle to represent portions of the earth in different months, and label them as in Fig. 16. Now place a lamp in the center to represent the sun as a source of light, and represent the earth by a small globe with a stand (Fig. 4). Have the flame at the height of the middle of the globe. First place the globe at the winter solstice (December 22d) and incline its upper (north) pole away from the sun, with the axis at the angle of $23\frac{1}{2}°$ Note the part of the room toward which the axis inclines. Keeping the axis always parallel to its original position, move the globe to different parts of the orbit in succession, and study the distribution of light on the globe. Turn the globe on its axis and observe the phenomena of long and short days in different latitudes. Follow all the explanations of the section.

FIG. 4.—Method of illustrating changes of the seasons.

21*a*. **Observation of the Sun's Position.**—It will be well if this exercise can be carried on once a week from December 22d to June 21st. The observations will need to be made largely at home, and will require some perseverance on the part of the student. Mark the time of sunrise and record it in tabular form as below. With a

THE SUN.

Date.	RISES.		Noon. Altitude.	SETS.	
	Hour.	Position.		Hour.	Position.

compass determine the position on the eastern horizon at which the sun appears and record. If a morning is cloudy, take the next morning, noting in the record the change from regular date. If there is a succession of cloudy days, omit the week's observations. Make similar records of time and position of setting.

In the school building, if possible, or at home, select a window facing exactly south.* On the side of this window fasten a quadrant, as in Fig. 5 of this manual, with mn horizontal. Let there be a projecting nail or peg at m. Let x represent the rays of the sun entering the side of the window at m at exact noon. Note the point (here 40°) on which the shadow of the nail falls. The angle b by

* In this exercise determine the true rather than the magnetic south (see Exercise 237), and take account of the difference between standard and local time

geometry = angle a, which is the noon height of the sun for this date. Record this each week.

How much change in the sunrise have we in one month? In six months? Difference in length of day between December and June solstices? Total change in noon altitude of the sun? Total change of direction of rising or setting?

24. **Time.**—One hour of difference in local time corresponding to 15° difference in longitude; what time dif-

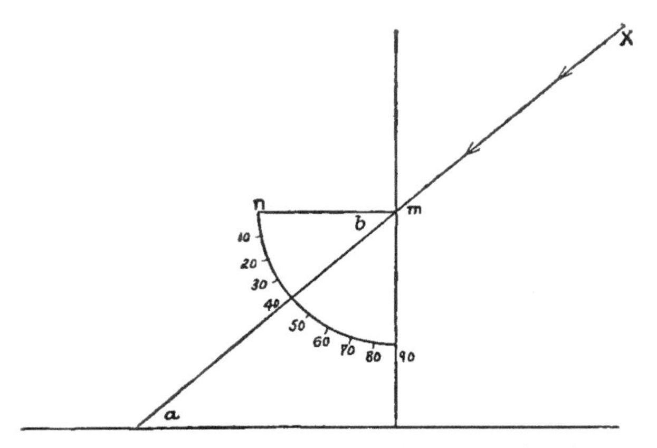

FIG. 5.—Method of finding the sun's noon altitude.

ference corresponds to 1°; to 1′; to 1″; to 50°; to 130°? On an outline map of the United States plot the standard time belts (see section 24 for data). The belts should be shown by use of colored crayons. What is the difference in local time between points differing in longitude by 22° 10′ 15″? What is the difference in local time between Portland, Me., and Portland, Ore.? What is the difference in standard time? What is the difference in longitude of the places differing 17′ 43″ in local time? The longitude of Washington is 77° 00′ 33.5″ W.; is the local time faster or slower than standard time (Eastern district), and how much? How much does local time differ from standard time at Indianapolis? At Chicago? At Denver? In San Francisco local time is ten minutes slower than standard time (Pacific); what is the longitude of San Francisco?

What is the difference in time between New York and Berlin? If a cable message is sent at 3 P. M. without delay from the former to the latter, at what time is it received? If sent from Berlin to New York at 3 P. M., at what hour is it received? If a traveler left Liverpool at three o'clock March 3d, and arrived in New York at three o'clock March 10th, what is the actual length of time consumed in his voyage? What experience would he have with his watch during the voyage? If he went around the earth eastward during exactly 100 complete rotations of the earth, how many days' doings would be recorded in his diary? Suppose he went around in like manner to the westward? The meridian 180° from Greenwich is agreed upon as the international date line, at which each day in the calendar is considered to begin and end. When vessels cross this line, the sailor drops a day if going eastward, and adds a day when moving westward across the line, to keep his reckoning right. Any other meridian might have been chosen, as with the prime meridian at Greenwich, uniformity being the chief matter of importance. It is, however, less inconvenient to employ a line which mainly lies across uninhabited seas.

CHAPTER III

RIVERS

27. **Field Study of a Gorge.**—How deep is it? Determine first by estimate. Compare with the height of a tree growing at the bottom or on the slope at either side. Or compare with the height of a house or church tower. The height of a man is a good unit of measurement, if the gorge is not too deep. The depth may also be found by the use of a hand level. The method is given in the Teacher's Guide, page 27.

Sides of the gorge. Do they show vertical walls, or slopes, or both? If both, how are the two related to each other? What is the angle of slope? Get your result first by estimate. Then use the clinometer. Method is given in Teacher's Guide, page 28. Compare the walls in slope and general appearance with the following shown in the text-book—frontispiece, Figs. 17, 18, 45, 51, 52, and 64. What causes the alternation of slope and cliff in Fig. 64? See pages 89 and 91 of text-book.

What is the material in the walls? If loose, is it fine or coarse? If stony, are the stones angular or rounded? Have any of the stones come down from rocky cliffs or ledges above? Do the cliffs overhang at any point? Can you see any blocks that are nearly ready to fall? What forces will push them off their perch? Will such changes go on more actively at certain seasons? If this continues for a long time, will there be any change in the form of the gorge? If the walls are of rock, what is its color and general appearance? Is the rock in layers? Are they thick or thin, solid or crumbling? Compare Figs. 17 and 52. If a talus or waste slope lies below a wall of rock, what is the relative height of the two? Do the tops of the gorge walls meet the adjoining surfaces of the fields by an angle, or pass into them by a curve?

The bottom of the gorge and the stream. Is the bottom wide or narrow? If the gorge is cut into solid rock, does the stream flow on a rock floor, or is this floor covered with waste? If we think of the stream as a carrier, is it keeping up with its work or not? Suppose the stream to be flowing on a rock floor. Can you show that the gorge is becoming deeper? In what ways is the rock being worn or cut away? Effect of stream, joints, frosts, roots, ice? If there are potholes, describe them. Is the waste in the stream bed round or angular? If the former, have the fragments been worn by rolling in the stream or were they rounded before they were seized by the stream?

As you go up the stream, is the ascent uniform, or by a series of rapids and reaches? Are there waterfalls? (These will be the subject of a separate exercise.) Make a longitudinal profile of the bed of the stream. Make a cross profile, using the elements of depth, slope, and width already determined. Is the valley V-shaped, or U-shaped, or would neither of these letters suggest its form?

27*a*. **Map Study of a Gorge.**—Kaibab Sheet, Arizona. —For account of the region, see pages 71 and 175, with frontispiece and Figs. 50, 64, and 126. Also Use of Governmental Maps in Schools, Davis, King & Collie, page 16. Plot the outline of the sheet on a general map. What are the scale and contour interval? What are the latitude and longitude of the quadrangle? Describe the course of the Colorado River across the area. How many tributaries are shown, on the north, and on the south? Why so few? What is the general altitude of the plateau? How deep is the canyon sunk below the general level?

How high are the walls of the outer canyon? See opposite L of Colorado River. How wide is the outer gorge at this point? How deep is the river gorge here?

Make a north and south profile of the canyon through L of Colorado River. Use horizontal scale of map, $\frac{1}{4}$ inch = 1 mile. Vertical scale, $\frac{1}{8}$ inch = 1,000 feet. What is the approximate vertical exaggeration? Compare your profile with the combined section and profile of Fig. 50. As directed below Fig. 50, see what parts of the gorge are shown in frontispiece and in Fig. 64.

How wide a belt along the river from east to west shows the plateau well cut into fragments and spurs? How do Kanab and Cataract Canyons compare in depth with the Colorado?

27*b*. **Dynamic Study of a Gorge.**—This study will commonly be made only in a laboratory which is fitted with a tank and hose. Make a heap of sand with gentle slopes and irregular surface and turn on a gentle spray. Note the absorption of some of the water, and the gradual gathering of the rest along one or more lines of depression. Note the small canyons formed. What is the character of the slopes? If a few small stones are mixed with the sand, what effect will some of these have when the rivulet crosses them? Do you recognize in the experiment any features seen in a field visit to a gorge?

Build up a hill with alternate layers of variable thickness, of sand and of fine tough clay, finishing with sand at the top. Be careful to use a fine spray, lest the work be too violent for the soft materials. Note the results; in rate of sinking of the canyon; in the longitudinal profile, and in the cross profile, or character of the side walls.

If the above apparatus is not at hand, the interested student can try similar experiments at home, if there be a door-yard or garden, with hose, or a common water-ing-pot with a fine sprinkler can be used, and the student can exercise his ingenuity in imitating the conditions of nature. Other results than the excavation of canyons will be produced. Some of these will be noticed in later exercises.

28. **Transportation by Streams.**—If you have visited a gorge, and found running water there, you have witnessed transportation accomplished in the spring floods. As all streams carry loads, whether flowing in gorges or elsewhere, we give a separate exercise.

Observe the stream at a point where it is swift. What is the size of the materials lodged in its bed? If there were more, or swifter water, would some of these stones move on? What sort of material is moved, at some point where the current is gentle? Are there, by the stream or in its bed, masses that came down at high water? Will they grow smaller as they are bowled along by successive floods? Will any causes tend to split them as they are long exposed?

Why is the water of a brook darkened after a shower? Collect a bottle of this water and allow the earthy matter to settle. How thick is the layer of sediment at the bottom? Make a rough comparison of the bulk of the water and of the earthy matter carried by it. When you collect the sample of water, make at the same point an estimate of the average depth and width of the current. The product of these two will give you the area of the cross-section. Pace off 50 or 100 feet along the stream. Drop a chip at the upper limit and time its passage to the lower point. This marks the velocity. Find how far the chip would travel in an hour, the velocity being uniform. Multiply this distance by the area of the cross-section. The product will be the amount in cubic feet of the water passing your point of observation in one hour. Multiply this by the ratio of mud to water as shown in your bottle. The result is the amount of waste gathered from the land and carried past the given point in one hour. Find the amount for a day: for the period during which the flood is at about the same height.

31. **Map Study of Divides.**—Review the definition of a divide in the text. Take any map which shows the drainage of New York. Observe that all waters flowing to the north and west enter the lakes or the St. Lawrence. Follow the divide that separates these waters from those that flow southward. What well-known lake and river in southwestern New York discharge into the Mississippi? How far is the north end of that lake from Lake Erie? How do the Chemung and Chenango send their waters to the ocean? Find the line of water partings between the Hudson River system and Lake Ontario; the Hudson and St. Lawrence River; the Hudson and Lake Champlain. If an outline map is at hand, trace on it the line of divides thus far studied.

31*a*. Elmira Sheet. What is the principal stream? What larger river does it join? See above, or general map. Note the small stream flowing north from near Horseheads. What lake does it enter? See Watkins sheet. Observe the course of Newtown Creek. How near do Newtown Creek and Catharine Creek approach each other? See scale. What is the character of the surface between them? Observe that this is one point along the divide already traced. It is also of interest as the point of overflow of a large glacial lake.

31*b*. Oriskany Sheet. Follow the course of the Mohawk, first southward, and turning eastward at Rome. Then note the course of Wood Creek so far as seen on this sheet. If the Oneida sheet is at hand, follow it to Oneida Lake. How far from the river to the creek through the city of Rome? What is the character of the surface? How do the waters of the two streams reach the ocean?

What is the altitude of the divide? This is the famous "Oneida Carrying Place" used in early history, and by the Indians before the coming of the white man. It also had been the point of outlet of vast glacial lakes. See text, page 146 and Fig. 112.

In place of the above exercise, study the system of divides in your own State, using any maps that are available, and looking to the teacher for suggestions. Pennsylvania, West Virginia, North Carolina, Wisconsin, Minnesota, Wyoming and Colorado are well fitted for this kind of study.

31*c*. **Divides Having Sharp Crests.**—Leadville Sheet. What is the chief stream in this quadrangle? Where is Leadville with reference to it? Answer the above in regard to altitude and distance as well as direction. What mountain-range east of Leadville? Note its altitude and its height above the Arkansas River. What stream drains the eastern slope of this range? What range west of the Arkansas River? Note the head-waters on the west edge of the map south of Hagerman Tunnel. They belong to Roaring Fork. On a general map learn the course, toward the sea, of Roaring Fork, Eagle River, Ten Mile Creek, and Blue River. Study the divides between these and the Arkansas and South Platte. Make an east and west profile along the line of Sacramento Gulch, Mount Sherman, Malta and Evergreen Lakes. Take horizontal scale of map and vertical scale, $\frac{1}{4}$ inch = 1,000 feet. Let the base-line be sea level. Observe the altitude of Leadville relative to sea level and to the mountain crests. How great slopes per mile do you find on either side of Mount Sherman? Observe that the Sawatch Range is much higher at Mount Massive than where our profile is drawn; also that but little of the relief on the west is shown on this sheet. Follow the crest of Park Range from Weston Pass to Bald Mountain, and note the variations of altitude.

The student should observe that 25- and 50-foot as well as 100-foot contours are used on some parts of this sheet. The last only will be needed in this study.

32. **Waterfalls—Field Study of a Local Fall.**—Brief description of gorge and stream. Does the gorge extend above as well as below the fall? If not, why? What is the height? How does the width of the stream at ordinary volume compare with the width of the gorge? Is there a clear plunge, or otherwise? What are the appearance and structure of the rocks? How does the structure affect the character of the fall? Is there a hard bed at the top? If so, are the softer rocks below undercut? Compare Fig. 23. Are there evidences of recession, such as fallen blocks? Is the rock jointed, and if so, what is the effect upon the fall? Is there a pool at the base of the fall, and if so, can you determine its depth? Make a longitudinal profile and section. For a model see Fig. 25. If the rock is not stratified, remember that your diagram will differ considerably from that in the text.

32*a*. **Niagara.**—Niagara Falls Sheet. Observe the character and altitude of the surface from Lake Ontario to Lewiston and Queenston. Study the escarpment which runs east and west from these points. How high is it? How far from Niagara Falls? What kind of surface do we find south of the escarpment? How much do the banks rise above the river at Grand Island? Study the diagram, Fig. 25. What rocks predominate near the top? What nearer the bottom? What causes the undercutting? How does the height of the falls compare with the depth of water below the falls? Why is the water so deep? Note that on the map it is represented here as still. Compare the appearance of the rapids as represented above the falls or in the direction of the whirlpool. Where on the map do the falls appear to be rapidly receding? Where would you infer that the falls were at the beginning? Make a cross profile just below the falls; another below the railway bridges. Does Niagara differ in principle from the Gadsden fall, Fig. 23? What differences of condition, or of details, do you notice?

37–40. **Field Study of Flood-plains and Meanders.**—How wide is the flood-plain? Is the stream midway or nearer one side? Is the plain smooth, or interrupted by old channels? Are any parts of it higher than others? If higher along the stream, why? See text-book, page 47. How much water in the stream—that is, is it an average height, or low, or high? If you have seen it washing over the flood-plains, describe. Why would earth and coarser waste be left on the fields at such a time? How has the plain been built up?

How much does the stream vary from a direct course through its flood-plain? Make a plan or map of its curves, showing it in either direction as far as you can see distinctly. Sketch in also the outer borders of the flood-plain at the base of the valley sides. Make a cross profile of valley, including stream, plain, and slopes. Select a curve for study. How great is the sidewise swing which it makes? Does it curve so strongly as to make an ox-bow? What is the form of the bank on the outside of the bend? On the inside? Why? How deep is the water on the two sides of the stream? Will the spur within the bend grow longer? What will happen to the bluff on the other side? How will these changes affect the form or size of the meander? Can you find a place where a "cut-off" has taken place, or may soon take place? Can you find abandoned meander courses on the flood-plain? Do they show meadow, marsh, or lagoon? If you have seen a meandering stream much larger or smaller than this, how wide was the swing of its meanders, as compared with those seen here?

How do the fields on the flood-plains compare with

the hillsides, as respects homes and crops? Where are most of the pastures and forests? Why? Compare the soils with those of the uplands. If the valley is spacious, where are the railway and chief highways?

37–40*a*. **Map Study of Flood-plains and Meanders.**—Kansas City Sheet. Marshall Sheet, Missouri. Begin with the former and note scale and contour interval. Plot the boundary of the sheet on a general map. Altitude of the uplands adjoining the valley of the Missouri River? How much is the valley sunk below the general surface? How wide is the flood-plain at Leavenworth? Include the stream in the measurement. Width at Connor? How many times does the Missouri River cross the flood-plain from one valley side to the other within this area? Are there any lakes representing former sections of the river? Take the Platte River, not confusing this with the greater river of the same name. How does the swing of its meanders compare with that of the trunk river? Are there any cut-off lakes? How could the Platte have flowed so far on the Missouri flood-plain without joining the main river? Give a possible reason for its abandoning this channel and entering the Missouri at Waldron. Why should the chief city have grown up where Kansas City now is? Note the railway routes.

Study the Marshall Sheet in a similar way. How far down the river from Kansas City is the Marshall quadrangle? How wide is the flood-plain here? Is there any difference in the number of lakes showing abandoned parts of the old course? What feature in the eastern part of the area is different from anything seen on the Kansas City Sheet? Compare the meander swings of Grand River with those of the Missouri. Compare the swings of the lower course of Big Creek with both the others.

37–40*b*. St. Louis, East Sheet. How is the present course of the stream related to the flood-plains? How wide is the flood-plain at the north? At the south? How much is the flood-plain below the adjoining uplands? Make an outline sketch of the flood-plain, showing its borders, the course of the river, and the position and form of the principal cut-off lakes. What conditions would serve to locate the city where it is? This question has reference only to this quadrangle, not to general geographic conditions.

42. **Dynamic Study of Deltas.**—Tank and spray as in section 27*b*. Fill the bottom of the tank so that your miniature land has water around it, or at least on one side. Use the fine spray to form a stream system as before, but watch now what goes on at the edge of the water. Record the progress of growth of the delta. If the material of the land is compact and the spray very gentle, the work may go on for several days, and notes be made each day or half day. What is the shape of the delta—that is, its outline—particularly in front? If lobate, do some lobes grow faster than others? How steep are the frontal slopes? Would the slope depend on the size of the fragments of waste? What causes a lobate growth of the front? Does the top surface of the delta show channels leading to the lobate points of discharge? Observe the progressive development of the valleys while the delta is in process of formation. Compare your delta with those shown in Figs. 37 and 40.

Draw down the water so that the upper half or more of the delta is exposed, and continue the spray. What effect is produced on the new-made land? Will a new delta form, and at what level?

After the material has partially dried, make a longitudinal section of your delta and see if the structure is similar to that shown in Fig. 39.

42*a*. **The Delta of the Nile.**—Read any available descriptions of the great river and its delta. See any encyclopedia, articles Egypt and Nile; also, Lyell, Principles of Geology, eleventh edition, vol. i, pp. 427–435.

Turn to the map, Fig. 38. The scale of this map is about 50 miles to the inch. The head of the delta is at Cairo. How far is this from the sea, directly to the north? The base of the delta may be said to extend from Alexandria to Port Said. What is the distance between the two points? How many *large* branches or distributaries run from below Cairo to the sea? At what town does the more easterly of these reach the Mediterranean? At the mouth of the other, northeast of Alexandria, is Rosetta, not indicated on this map. From Damietta to Rosetta is sometimes considered the base of the delta. What is the cause of the capes at these two points? The lower portion of the delta has undergone gradual subsidence. What feature of the northern parts of the area would be explained by this? Is there anywhere else in Egypt so large a body of fertile land? Consult a map on which the towns are fully shown, as in the Century Atlas, and record your opinion of the density of population.

48. **Map Study of Lake Filling.**—Fig. 42, or the Watkins, N. Y., Sheet, entire. We will first consider what we have in Fig. 42. What is the altitude of the surface of Seneca Lake? See near top of map. What altitude is marked by the contour adjacent to the lake? Do any contours cross the area between Watkins and Montour Falls? How far up the stream from Montour Falls before you cross a contour? How far is this point from the lake? How wide is this flat area? How many square miles does it include approximately? Note Watkins Glen and lesser gorges of other streams which open on the flat ground. Can you explain the conditions otherwise than by a process of lake filling? Observe that we have here a delta, though not one of salient form, because the waste is deposited in the head of the lake in a narrow

valley. The lake is but a few feet deep about the entrance of Catharine Creek; about 50 feet deep adjacent to Watkins, but becomes from 150 to 200 feet deep within about one-half mile from the south shore. May Catharine Creek have flowed into the lake at other points? Note Salt Point, and explain. If the Watkins Sheet is at hand, observe the extent of Watkins Glen and others, as affording material for filling. Note other points below Salt Point; also Glenora and Peach Orchard. Combine this sheet with the Elmira, already studied, and trace Catharine Creek up to the divide.

45–50. **Comparative Study of the Maps, Figs. 41 and 44.**—Mature and imperfect dissection by streams. Fig. 41 is part of the Huntington Sheet, West Virginia. Note the scale and contour interval. How many square miles are shown by the map? What is the chief stream? This enters the Ohio River a little more than 20 miles to the northwest. What is the altitude of Mann Knob? What is the average altitude of the hilltops throughout the area? How far above the sea level is the floor of the Guyandot Valley? Are the branches of this river many or few? Have the streams any considerable flood-plains? How would you describe the surface of this region as a whole? Do the roads run on the uplands or along the streams? If a copy of the Huntington Sheet is at hand, extend your study over the entire area, with questions similar to those given above. In this case record the position of towns and smaller settlements, as regards the streams. Study section 45 in connection with the above exercise.

Fig. 44 is part of the Piney Point Sheet, Virginia. Note scale and contour interval. Observe that the interval is but one-fifth as great as in Fig. 41. Why is a small interval needed in a flat country? Observe also the difference of scale as compared with Fig. 41. The water at the northeast is a bit of the Potomac River. Does this differ from sea level along the lower river? Can you show this from the map? Are the streams here many or few? How wide are the smooth uplands between the heads of the streams? What is the greatest altitude you can find here? How do the roads run as regards the streams? Can you suggest a reason for this?

If you have the Piney Point Sheet study it in a similar way. Note the tidal mouths of streams both on the north and south of the Potomac; the numerous flat tabular lands between the streams; and the location of the settlements. This is a young plain imperfectly covered by the drainage system, or but partly dissected, as compared with the full possession by the streams seen in the West Virginia area. Read carefully section 50. It would be useful to read also in this connection section 150, describing the Atlantic Coastal Plain, of which this is a small part.

51. **Trellised or Rectangular Drainage.** — Monterey Sheet, Virginia and West Virginia. Folio 61, United States Geological Survey (price 25 cents), gives in detail the topography and geology of this quadrangle. It may be studied with profit, though not essential to this exercise.

In which State does the larger part of this area lie? Does the line follow a mountain ridge or a valley? Note the scale and contour interval. In what direction do the chief streams run? Determine from a general map to what larger rivers they are tributary. What are the directions of the smaller branches of these streams? What is the elevation of the mountain ridges above the main valleys? Do the larger streams anywhere cut across or through the ridges? See Black Creek, Little Black Creek, and Bull Pasture River. Note the mountain ridge east of Greenbrier River. Why should it bear six names within a distance of 20 miles? Observe in like manner the ridge east of Jackson River. With tracing paper make a plan of the drainage of the northeast quarter of the map and note the rectangular arrangement. Find the only bit of railway in the area, and observe that it runs through a "gap." How will the form of the ridges be affected when the streamlets have cut deeper and longer channels?

Compare the drainage shown on Pine Grove Sheet, Pennsylvania. Observe the relations of this area to the Susquehanna River. What is the chief stream in the quadrangle? Find Tremont. How many "elbows" along the stream before it leaves the quadrangle? Name the mountain ridges which it passes by water gaps. What is the direction of flow in these gaps? Such parts of a stream are often called "transverse." What is the direction of the stream when following the valleys? Such parts of a stream are often called "longitudinal." Find other transverse and longitudinal sections of streams. Note the tributaries of Black Creek as to size and number. Trace the drainage of that part of the map lying south of Tremont.

The student should give special attention to section 51, and to Fig. 45, showing a water-gap, and Figs. 46 and 47. Figs. 44 and 41 also show the branching or tree-like drainage in contrast with that just studied. The Harrisburg sheet shows a series of great water-gaps occupied by the Susquehanna. Observe that it is the master streams which have been able to sink their beds across the hard ribs of rock encountered in the down-wear of the land, and as fast as the cutting of the gaps permitted, the softer beds have been cut away, forming the valleys which lie between the ridges.

54–62. **Outline Sketches of River Systems in the United States.**—Rivers in blue, divides in red. Upon an outline map trace the chief rivers of the Atlantic slope. (This slope in a limited sense, not now including the St.

Lawrence or the Mississippi.) Trace the line of water-parting that separates these lesser rivers from the St. Lawrence and the Mississippi. In what part of New England does it run? We have already followed the line across New York (exercise 31). What river does it separate in Pennsylvania? What general course have all the Atlantic rivers south of the Hudson? South of the Potomac River, does the divide run nearer the east or the west border of the Appalachian Mountains? That is, does the mountain-belt drain chiefly into the Mississippi or directly into the Atlantic? (Consult any map of the United States which shows relief.) What proportion of Ohio, Indiana, or Illinois is tributary to the Great Lakes? What share of Minnesota (estimate from your sketch map) drains into Lake Superior? Into Hudson Bay? Into the Mississippi? Analyze in a similar manner the drainage of Colorado; Wyoming; Utah; Montana. Through how many channels does the Pacific slope send most of its waters to the sea? How does the Atlantic slope compare in this respect? What share of the United States drains into the Atlantic, using the words Atlantic slope now in their broadest sense?

55. **The Mississippi River.**—It is not expected that any class will complete all of the exercises that follow. The choice will depend on the available material, or on the location of the school, especially if within some part of the Mississippi basin. The material for study is given under each exercise. The following references will be useful: Hall's Geography of Minnesota (The H. W. Wilson Company, Minneapolis), pp. 118–125, 129–133, 153–157; Russell's Rivers of North America, pp. 70–75, 118–123, 265–271; Stanford's North America, vol. ii, Gannett, pp. 17–21, 400–402; International Geography, pp. 743, 748.

55*a*. **The Mississippi System.**—On an outline map of

the United States sketch the chief members of the system and the bounding divides. This in part will be a review of exercises 31 and 54–62. Of the greatest eastern, northern, and western members, the first (Ohio) discharges the most water, the second occupies a central position, and the third is the longest. Which in your opinion has the best right to be called the Mississippi River?

Construct profiles of the several chief parts of the system, using altitudes and distances as below: ½ inch = 100 miles is a suitable horizontal scale, and ½ inch = 1,000 feet may be used for a vertical scale. This will require special care in drawing where the descent is slight. If Gannett's Profiles of Rivers in the United States is at hand, some of the exercises can be carried to much greater detail by using intermediate stations. The pamphlet is No. 44, Water-Supply and Irrigation Papers, United States Geological Survey.

Data for profiles of the Mississippi River, including the greater branches.

MISSISSIPPI.

Station.	Distance from mouth.	Height above sea.
	Miles.	Feet.
Mouth	0	00
Vicksburg	487	48
Memphis	862	185
Cairo	1097	274
St. Louis	1270	380
Mouth of Missouri River	1288	395
Burlington, Iowa	1492	511
Lacrosse, Wis.	1791	628
St. Paul	1937	683
Minneapolis	1952	794
Lake Itasca	2296	1462

Draw a straight line 23 half inches (or 11½ inches) long. Using the adopted scale, mark the position of Vicksburg (5 half inches), Memphis (about $8\frac{1}{2}$ half inches), Cairo, etc. Raise a vertical at Vicksburg 48-1000, or about 1-20 of a half inch in length. This will be scarcely more than a point, but it emphasizes the sluggishness of the lower river. At Cairo the vertical will be a little more than $\frac{1}{8}$ inch, at Minneapolis a little more than $\frac{3}{8}$ inch, and at Lake Itasca about $\frac{3}{4}$ inch. Connect the summits of the verticals with a broken line, and you will have the desired profile. What is the meaning of the sudden change of height above St. Paul? See 55*b*.

What is the descent per mile between Vicksburg and the sea? Between Cairo and Vicksburg? Between Minneapolis and Lake Itasca?

OHIO-ALLEGHENY.

STATION.	Distance from mouth.	Height above sea.
	Miles.	Feet.
Mouth	0	274
New Albany, Ind	374	367
Louisville	378	394
Cincinnati	511	431
Wheeling	871	622
Pittsburg	963	702
Source of Allegheny	1300	1700

MISSOURI.

STATION.	Distance from mouth.	Height above sea.
Mouth	0	395
Kansas City	391	716
Omaha	660	960
Yankton	898	1161
Mouth of the Yellowstone, N. D	1549	1855
Fort Benton, Mont	2074	2565
Great Falls	2111	3295
Three Forks	2340	4000
Head of Gallatin River	2428	7000

By combining the above data and using great care an interesting composite profile of the three great sources may be drawn, which will show one line for the lower river, diverging above. What other branches of the Mississippi may be fairly compared with those studied above?

55*b*. **Mississippi River at the "Twin Cities."**—Minneapolis and St. Paul Sheets. Locate these on a general map. Note scale and contour interval. What changes of direction are made by the river as it passes the two cities? What is the length of its course across the combined quadrangles? Where is St. Anthony Falls? What is the altitude of the water surface above and below the falls? What is the character of the valley below the falls? Depth? Width? This gorge has been cut as the falls have receded, during a period variously estimated at from eight thousand to twenty thousand years. Compare Niagara. Note Minnehaha Falls and gorge, St. Paul Sheet. Why has this fall receded so slowly, making so short a gorge? What river joins the Mississippi at the foot of the gorge? Compare its valley with the valley of the Mississippi below Pike Island in the following particulars: width, depth, lakes, and swamps. Observe the lakes, swamps, and rough lands so common in the area. Compare exercise 134–141. Also see section 146 in the text for reference to the origin of this gorge. Consult also Hall's Geography of Minnesota, pp. 129–133. St. Paul is at the head of navigation on the Mississippi. What is the bearing of this, and of the existence of St. Anthony Falls, on the development of cities here?

55*c*. **The Mississippi Between Illinois and Iowa.**—Savanna and Clinton Sheets, Iowa and Illinois. Locate these on the general map. Which quadrangle lies to the north? What is the average width of the flood-plains in the two areas taken together? How wide are flood-plains and river combined west of Savanna? Can you find a contour line crossing the river? What does this mean? Can you tell from the map, in any other way, that the current is sluggish? (The fall at Clinton is 0.2 foot per mile and at Savanna 0.3 foot.) Where along the river are there steep bluffs? What is the cause of them? What reason for believing that this work will not be continued so long as man controls the region? Note the swamp about Dyson Lake and trace it to the Clinton Sheet. What is its origin? Explain the hooked lake west of Rush Creek. Do the waters of Rush Creek drain into the Mississippi, and if so, how? What is the average height of the uplands above the flood-plain? Is the upland smooth or rough? Compare the La Salle area (Fig. 11). Observe the roadways on the Savanna sheet. Do they run mainly in the valleys or along the divides? Account for their position as you find them. Make a cross profile of the valley on an east and west line crossing the hilltop in Savanna. Carry it eastward across the valley of Plum River. Use the horizontal scale of the map; vertical scale $\frac{1}{4}$ inch = 100 feet.

55d. **The Lower Mississippi** (from the mouth of the Ohio to the Gulf).—Map of the Alluvial Valley of the Mississippi River from the head of St. Lawrence Basin to the Gulf of Mexico.

By whom is this map published? What is the scale? Using this scale, determine the distance from Cairo to the mouth of the river. What is the distance along the river? See *55a* in "data for profiles." What States are touched by this section of the river? How are the flood-plain areas distinguished? How wide is the flood-plain at Natchez? At Vicksburg? At Memphis? What is the average width? How wide is the belt actually occupied by the river with its meanders? (Draw, or imagine, lines tangent to the principal outswinging beds on the east and west, and estimate the average width of the strip lying between them.) With a tracing paper make a plan of the meanders and cut-off lakes between Natchez and Vicksburg. How do Natchez, Vicksburg, and Memphis stand with reference to the flood-plains?

The delta region. Record the position of Donaldsonville with reference to Baton Rouge and New Orleans. What great bayou leads from the river at Donaldsonville? What are the chief lakes of the delta? How is the higher and drier land situated with reference to the streams? Consult page 47 in the text-book. Compare the width of the flood-plain of the lower river with its width at Savanna and at St. Paul.

55*e*. **Meanders and Cut-offs.**—Map of Lower Mississippi River from the mouth of the Ohio River to the Head of the Passes, in 32 sheets, scale 1 inch = 1 mile. These maps were published by the Mississippi River Commission, 1881–97. Many of them were republished in 1900 with red overprint lines showing the changes that had taken place in a period of several years. Sheets 13, 14, 18, 19.

Take 13 and 14, hanging or spreading them in proper order. What great river enters the Mississippi near the north end of this area? What are shown by the heavy red lines? By the light red lines? By the dotted red lines? Date of cut-off opposite mouth of Arkansas River? Why is Monterey Landing (Sheet 13) shown in red? Why is this a better place for a landing than the inside of the bend? How has Island No. 76 changed in form? Observe the neck between Miller's Bend and Georgetown Bend, Sheet 14. How did its width in 1894 (see Note) compare with its width in 1881–82? What is likely to happen here? How much would the river be shortened? Would Utopia Landing be likely to survive? Would it take long for Rowdy Bend to become a lake? See Beulah Lake, Sheet 13. How far is it from the mouth of the Arkansas River to Greenville direct? How far by the river?

Sheets 18–19. What important town at the north end of this area? When did a cut-off occur just below this city? Note also the name of the lake that was formed. Observe "Grant's Canal" south of the cut-off. In 1863 Grant tried to turn the river across the neck, so that his gunboats could go up and down the river without facing the guns of Vicksburg. What he was unable to do, nature accomplished thirteen years later.

Note Newtown Landing, Sheet 18. How far from the river bank is the position of the old upper Newtown Landing? How long between the earlier and later surveys? Note the bend at the northeast, Sheet 19. Does this bend lie north or south of its former position? How much? Observe also the bend about Rodney Island at the south end of this area. In which direction has it shifted? In like manner note the southward encroachment west of Newtown Landing, Sheet 18. Do these facts suggest a law as to the down-stream shifting of meanders? Examine other sheets of this series for similar cases, and find, if possible, a reason for such changes.

55*f*. **The Delta.**—Donaldsonville Sheet. The location has been noted in 55*d*. What is the contour interval? What is the highest contour on the map? Would the common 20-foot interval be of any service in mapping such an area? What share of the quadrangle is marsh? Where are the dry levels? How wide is the belt on either side of the river? In what way does the land slope in reference to the river? What is the cause of this? Where are the principal highways and most of the dwellings? Why? How do the minor roads run? Why? What are the blue lines? See "Legend." What part of the map shows on a small scale the features shown by the river? Find Nita Crevasse. Observe the alluvial deposit to the northward. Why has it a series of lobate forms? Make a profile of river, natural levees, and swamps along a line passing through the W of Winchester and the W of Whitehall. Use horizontal scale of map, and vertical scale $\frac{1}{16}$ inch = 5 feet. Compare this profile with those of Savanna and St. Paul.

If the Thibodeaux Sheet is at hand, note the repetition of the above conditions along Bayou Lafourche. In like manner Houma Sheet shows another section of Bayou Lafourche, with narrowing natural levees, while about Houma are several small parallel bayous and narrow belts of habitable land. Here, too, we see Field and Long Lakes, types of many lakes in the region, enclosed by the network of higher land along the bayous or "distributaries."

55*g*. **Mouth of the River.**—East Delta Sheet and West Delta Sheet. The former may be taken by itself. What is the scale? Are contours used? Why? See statement at bottom of sheet relative to altitude. Compare the conditions with those about Donaldsonville. Are there any houses, highways, or tillable lands? Where are the lands situated? With what areas on the Donaldsonville map do they correspond? What here correspond to the swamps about Donaldsonville? How many "passes" are shown on this map? Where are the jetties, and what is their purpose?

If West Delta Sheet is available, note South Pass and the minor passes to the westward. Combine the two sheets and observe the finger-like lands forming a fan-shaped pattern east of Cubit Gap. What has happened here? Compare conditions north of Nita Crevasse, 55*f*.

55*h*. **The Tributaries from the East.**—The details of this exercise must be left largely to the teacher and the student, according to local needs. The Wisconsin and Illinois Rivers may be studied in detail in the States to which they belong. Longitudinal profiles should be drawn (see Gannett as above for data). In all such cases we

FIG. 6.—Channels of the Platte River, near the Colorado line.

may ask whether we have a graded channel, nearing base level, or one which lacks much of grade—that is, shows waterfalls, rapids, and torrents. The lower section may be at grade, and the rest may not have reached this stage. Note also glacial interruptions, section 146. The Ohio system may afford many studies. For the trunk stream study: Cincinnati, East and West Sheets; Huntington, W. Va., Sheet (see Fig. 41, and exercises 45–50); also Wheeling Sheet, and, when published, the sheets for Pittsburg and Allegheny. For the Tennessee, take the Chattanooga Sheet, in connection with Fig. 133.

55*i*. **The Tributaries from the West.**—A profile of the Mississippi has already been made (55*a*). The Missouri, or its greatest branch the Platte, the Arkansas, or the Red may be chosen for intensive study, according to locality. For types of country drained see as follows: Fig. 18, Yellowstone Cañon; Fig. 123, Rocky Mountains in Colorado; Figs. 118, 119, 67, lands of the Great Plains.

Observe Fig. 6 of this manual. What is the river? At what point? Is the stream flowing to right or left? How can you tell? What evidence of fluctuation in volume? The Platte descends but about 4,000 feet in passing from the foot of the mountains in Colorado and Wyoming to the Missouri below Omaha. The soil and rocks are porous and the climate dry. Can you state how these conditions would operate to produce the results shown in the view? (See reference to overloaded stream, p. 34 of the text-book.) See also pp. 163, 320, 324 and Herbertson's Descriptive Geography, North America, p. 144. Observe the bluffs along the river. What do you say of their height and materials? Does the river appear to have cut effectively into the plain? It would be very useful to study here Wood River and Lexington Sheets, Nebraska. Observe "braided" channels, width of flood-plain, height of bluffs, sand-dunes from river sands (the last on Wood River Sheet).

55*j*. **Man's Uses of the Mississippi.**—Many suggestions have already been made. We may, if desired, take three heads, as follows:

Navigation.—The following is Gannett's table (Stanford's North America, vol. ii, pp. 401, 402). The depth chosen is 3 feet. For craft drawing more water the distances would usually be less. On your outline map of the system traverse the several streams with red ink to the limit of navigation. What is the total mileage?

NAVIGABLE WATERS

Main Mississippi to Cairo, Ill	650
Upper Mississippi to Minneapolis, Minn	700
Bayou Lafourche, La	100
Bayou Atchafalaya, La	150
Red River and branches, La. to Gainesville, Tex	600
Ouachita and branches, La. and Ark	450
Yazoo, Miss	300
Arkansas and branches, Ark. to Wichita, Kan	600
St. Francis and branches, Ark	300
White River and branches, Ark	400
Ohio River, to Pittsburg, Pa	800
Allegheny, Pa	125
Monongahela, Pa	100
Tennessee, Ky. and Tenn	550
Cumberland, Ky	350
Green, Ky	100
Kentucky, Ky	125
Licking, Ky	125
Big Sandy, Ky	125
Guyandotte, W. Va	75
Kanawha, W. Va	150
Wabash, Ind	150
Muskingum, Ohio	75
Illinois	200
Missouri to Fort Benton	1500
Gasconade	50
Osage	100
Yellowstone	250

Floods.—Note the rainfall of the Upper Mississippi, 35 inches; Ohio basin, 43 inches; Missouri basin, 20 inches. Which stream would contribute most to the floods of the lower river? How much do floods sometimes raise the water at Cincinnati? See text, page 48. There is sometimes a rise of equal amount at Cairo and Vicksburg and of 20 feet at New Orleans. See Figs. 29 and 30 in text. The United States Weather Bureau maintains many stations for flood warnings in the Ohio basin. How would these warnings be of use to the population between Cairo and the Gulf? Could a levee system ever be made a perfect protection? Does such a river "aggrade" or build up its bed? How would this affect the levee problem?

Cities.—These may be studied as fully as time permits as regards geographic causes for origin or growth.

The Home River.*—By this is meant the nearest considerable river. It may be a small river, soon entering the sea, or it may be a branch of a great river system. In the latter case carry your study to its junction with the main stream, or study it so far as it is found in your own State. If the altitudes are available, make a profile of its bed from its source. On an outline State map sketch it in some detail. What is the form of the valley so far as you know or can learn by inquiry? If the contoured maps are at hand, you can learn this very definitely. Are there rapids or waterfalls, and what is the course of them? If the valley is wide in some places and gorge-like in others, note this fact and watch for an explanation to come later in your course (section 146). If the annual reports of the United States Geological Survey are at hand, see if you can find studies of the amount of water flowing in your river in the volumes devoted to hydrography. Are there bottom lands, and are they valuable for tillage? Is the stream navigable? Are there towns on its course, and how closely do they depend on the river for their origin and growth?

---

* This and a few other exercises are not numbered, since they do not relate to specific sections in the text.

CHAPTER IV

WEATHERING AND SOILS

Reference is made in the text-book to several of the chief kinds of rock. As all rocks are made up of minerals, we will first select a few of these substances for study. There are hundreds of kinds of minerals, but rocks are mainly composed of a few. Familiarity with half a dozen of these is all that is needed in physical geography.

Quartz.—Specimens of glassy quartz and of flint. Take first the glassy quartz. Is the mineral transparent? Is it in crystalline form? If so, how many sides has the crystal? If irregular, does it break in any special direction, or as if by accident? How hard is it? Will glass scratch it? Can you scratch it with the point of a steel knife? Try dilute hydrochloric acid on it. Does it show any effect? What would you say of the endurance of a rock chiefly composed of it? Test your flint in the same way as to hardness and solubility. Is it transparent? Translucent? Opaque? How does it break? If in smooth curves, this is called "conchoidal," or shell-like fracture. What uses have been made of flint, depending on the way in which it breaks or flakes? There are other varieties of quartz. Among those are milky quartz, smoky quartz, amethyst, rose quartz, agate, and jasper. All these varieties depend chiefly on the presence of other substances which change the color and general appearance.

Feldspar.—What is the color of your specimen? Does it break irregularly or along certain planes? (The property of splitting along parallel planes is called cleavage.) Does the feldspar cleave in one, or two directions? If more than one, give the angle between the two. Will glass scratch the specimen? Will it scratch glass? Test it in this manner with the quartz. What is the order of hardness of these three? Not all feldspars will have the color of your specimen. They may be white, gray, green, yellow, or light red. They break up or disintegrate on exposure much more easily than quartz. See under granite and shale, 67 and 65.

Mica.—What is the color of your specimen? Is it transparent? Translucent? If the latter, would a thinner specimen be transparent? Is the cleavage perfect or otherwise? Has mica the property of elasticity? Test with thin sheets. Has your specimen crystalline form? If you have ragged or clipped sheets this will not appear. If you find crystalline form, what is it? How hard is the mica? Test with a piece of glass; with your finger nail; with a piece of gypsum. The common mica is called muscovite, or sometimes Muschovy glass, from its use for windows in parts of Russia. It is incorrect to call it "isinglass."

Calcite.—The student may best begin with a fragment of Iceland spar. Test its hardness; with quartz; feldspar; mica. How does it compare with each of these? What is its color? Capacity for transmitting light? In how many directions does it cleave, and at about what angles? What happens if dilute hydrochloric acid is applied to it? This is not the only crystalline form of calcite. Sharp-pointed crystals, known as dog-tooth spar, are common. Much calcite is not in the crystalline form. If a piece of native chalk is at hand, note its color, texture, hardness, and the effect of acid. Calcite is one of the most common substances, as in limestones (66), shells, corals, and bones.

Gypsum.—Begin with a fragment of selenite. What

is its color? Luster? How does it compare in hardness with the crystalline calcite? How does its cleavage compare with that of mica? Is it affected by dilute acid? Compare any piece of massive gypsum, if at hand. Gypsum is a sulfate of calcium, being composed of calcium, sulfur, oxygen, and water; while calcite is a carbonate of calcium, containing carbon, calcium, and oxygen. When the water of gypsum is driven off by burning, it becomes "plaster of paris."

The Iron Compounds.—*Hematite.*—Is the specimen crystalline or non-crystalline? What is its color? If black, reduce a fragment to fine powder. What is the color of the powder? Will a magnet attract bits of it? Study the magnetite in the same way. What differences do you find? Study the limonite if a specimen is at hand; also iron pyrite. What is the color of the last? Its luster? Its crystalline form? With what metal has it sometimes been confused? Observe that the hematite and magnetite are the most important ores of iron, while the pyrite is not used for the extraction of iron, but sometimes carries gold, and, containing sulfur, is sometimes employed in making sulfuric acid. Iron makes but a small part of the earth's crust, but is widely diffused, and has much to do with the color of rocks.

64. **Sandstone.**—Take any specimen of sand from the nearest sand-bed, or from a lake shore or marine beach. Place a pinch of the sand under a hand magnifier, or a low-power microscope. Observe the shape of the grains, whether angular, or worn and battered. Note the proportion of quartz and of other minerals. Next take specimens of sandstone. Note the color and texture, whether coarse or fine. If the former, are any pebbles present? If so, it may be called conglomerate, or puddingstone. Crush a small fragment and compare the resulting grains with the sand already studied. Do you find any encrusta-

tion on the grains that may have served as a cement? Test with acid to see if calcareous cement is present. Judging by the cement, or by the difficulty of crushing, which do you think would be the most durable of your specimens in a wall or a natural ledge? Drop a little water on the specimen, and note whether it is absorbed slowly or rapidly. Are the color and constitution the same throughout the specimen? Does it show bands or layers? If so, what is their meaning?

65. **Shale.**—What is the color? Can you detect the grains of which the rock is made up? How readily does the rock split? What is the thickness of the resulting flakes? Observe that these planes of splitting are due to stratification, and are not the same as the cleavage planes of minerals. Compare all the specimens of shale which you have, as regards color, feeling, and general appearance. Make similar notes upon any specimens of clay from brick-yards or elsewhere. Compare sun-dried fragments of clay with the shales. Have you a rock intermediate between shale and sandstone? Is it gritty to the touch? Does it seem to be made of distinct grains like the ordinary sandstone?

66. **Limestone.**—What is the color of your specimen? Note the texture, whether crystalline or fine grained and compact. Do any fossils appear? Test with dilute acid. If action is slow, crush a little of the rock for the test. Allow the powder to remain until all possible solution has taken place. If enough acid has been used, the residue will be of other minerals present in the stone. If several specimens are at hand, compare them as to the above points. Limestone may be subjected to pressure and other forces which change its texture, color, and general appearance, so that we call it marble.

67. **Granite.**—See section 67 and Fig. 53. Observe your specimen, looking for the separate minerals which

form it. What are the glassy bits? Do they show any crystalline form? Are they mingled with the other minerals in any particular manner? Do you see bits which are white, green, pink, or yellow? Do they show cleavage faces? What is the mineral? Do you find small spangles of a shiny mineral? What is it? These three make up a typical granite. If you have several kinds of granite, compare them as to color and the coarseness or fineness of the mineral bits of which they are made up. Which mineral gives the prevailing color to the granite? Which would resist exposure best, a rock of one mineral or one composed of many pieces of several minerals packed together? Which mineral of the granite decays rather readily? (Section 67.) How many kinds of granite have you seen, in buildings, burial monuments, and any other structures, near your home?

68–76. **Field Study of Weathering.**—Visit any cemetery. What is the appearance of marbles that have been set within three or four years? Find one which has been set fifteen years. Has it lost its polish in whole or in part? What is the cause? Find marbles that have been exposed thirty or forty years. Do any seams appear? Have the corners scaled anywhere? Are there lichens growing in the grooves of the lettering or on the general surface? The marbles often have a limestone pedestal. Examine these blocks. Have the bedding planes come to light? Is the top surface roughened and pitted by the solvent work of rain-water? If you have access to an old cemetery, observe the blocks of slate which have stood for two hundred or three hundred years, as in the old burial places of Boston. Would marbles have stood as well?

Note carefully the door-stones, water-tables, window-bases, and walls of buildings erected wholly or in part of stone. Study the carved work of approaches to brownstone dwellings in New York or other cities, and the pillars, pilasters, or exposed cornices of churches and other public buildings. Where several or many structures have from time to time been erected of the same stone, note the differences of condition, particularly the progressive changes of color that have taken place. In this connection observe Fig. 57. It is called in the title an "old" building: so it would be if judged by the standards of the New World, being somewhat less than 300 years in age. But for Europe this is young. The rock dissolves and crumbles readily, thus giving a false appearance of great antiquity.

68–76*a*. **Field Study of Weathering.**—Ledges, quarries, and the drift. Visit a sand-pit or railway cut. Observe carefully the different kinds of material from the top to the bottom of the exposure. How thick is the soil? This can be roughly told by the darker color and the depth to which there is a thorough penetration by small roots. In the sand-pit do you find below the soil sand which is discolored? What is the color of the unweathered sand still farther down? In digging a ditch or well, the upper parts are often cut through reddish or yellowish material, while below the material is blue. What is the reason for this?

Observe the corners and faces of rock in a natural ledge or old quarry. Is there a broken, shaly appearance of the surfaces? Crack off a piece with a hammer. How does the appearance at the depth of two or more inches compare with that of the surface parts? Examine a heap of boulders or cobblestones. Can you find any that crush under the pressure of the hand or readily cut with a knife? Break some in two. Do you find in some cases a weathered rim and an unchanged core? Look for wedging by tree roots along the sides of ravines; also for blocks thrust somewhat from their natural position by water freezing along joint planes.

FIG. 7.—Toadstool Park, Sioux County, Nebraska.

68–76*b*. **Toadstool Park.**—Study Fig. 7 of this manual. In what State is Toadstool Park? Consult section 82 and Figs. 66 and 67. Observe in Toadstool Park the difference between the land forms of the foreground and distance. Which rocks are the hardest? Has the rain or the other processes of weathering had most to do in the foreground? In the background? What is the position of the strata in the foreground? Is there any difference in their hardness or resisting power? Are some hard beds thicker than others? Are some soft beds thicker than others? The vertical divisions seen here and there are joints (section 27). To what is the toadstool effect due? Has much or little rock waste been removed? Do you see any proof of occasional stream work in this view? Compare, for joints, Figs. 54, 55; for rain work, Fig. 56; for contrast as to talus, Fig. 64; and for bad-lands in general, Figs. 66, 67.

77. **Thickness of Land Waste.**—How far below the surface is the bed rock, on the average, in your region? Do you have any wells sunk to the level of the bed rock? What is the depth of the drift, if you are within the glaciated belt (section 142), or of the residual from rock decay? Other points for measures of thickness will be found in gorges, railway cuts, and where stripping has been done for quarrying.

77–79. **Collection of Soil and Drift.**—If desired, such a collection may be added to the school outfit. A half pint each may be gathered from soils of several fields, all sands and gravels available, muck and peat of swamps, clays, the finer waste of talus slopes, etc. Let the samples be enclosed in wide-necked bottles and labeled, stating material and locality. Let the student record in his notebook a short account of each specimen with the nature and origin of the material, so far as he understands these points.

89. **Water in Rocks.**—Take a specimen of sandstone and make sure that it is thoroughly dried throughout. Weigh it, and then thoroughly soak it in water. Weigh it again, and thus ascertain the percentage of water which the stone will absorb. Make the same test with other sandstones. What bearing does the capacity for absorbing water have upon the durability of building stones? The porosity of the sandstone illustrates the fact that sandstones form the reservoirs in which petroleum is contained (see section 173). Make similar absorption tests with limestone or granites, and compare the results with those obtained from sandstones.

CHAPTER V

WIND WORK

108–119. **Field Study of Blown Sand.**—What is the locality studied? What cause hinders vegetation from holding the sands down? What is the direction of the winds which transport the sand? What is the origin of the sand? What is the shape and composition of the sand grains as seen under a magnifier? How extensive is the area covered by the dunes? What is the height of individual hills? Are they irregularly assembled, or in wave-like ridges? What is the angle of slope of the dune surfaces? Have the dunes recently shifted position, as shown by destruction of trees or invasion of tilled fields? Is migration still in progress? How strong a wind is required to move the sands? How much plant growth, if any, on the dunes? Has any of it been caused by man to restrain migration? (Compare Fig. 86 and section 119.) Has any other means of checking the sands been used? (Compare Fig. 84.) Do the dune surfaces show ripple marks? Such may be imperfectly seen in Fig. 85. Sketch the profiles of a group of dunes.

112. **Map Study of Dunes.**—Kinsley Sheet, Kansas. Locate the sheet on a general map, and record scale and contour interval. What is the difference in the form of the contours showing dunes as compared with those representing surfaces shaped by the ordinary drainage? How are the dunes of this area related to the Arkansas River? Why? Lay out a plot 2 miles square and count the dune

hills shown on it. Can you tell exactly how high a given hill is? What is approximately the height of the most of them? How high are some of them? What reasons for exposure of the sands to wind action in this region? It would be useful to compare the Meade, Dodge, and Coldwater Sheets, Kansas, and the Kearney Sheet, Nebraska. Study the character of the Platte River and valley in relation to the belt of dunes. Compare 55*i* and Fig. 6 of this manual.

120. **The Sand Blast.**—In Fig. 87 where is the wearing by the sand taking place? Why at this level and nowhere else? What will happen in time if the process is continued? What other locality of such work is shown in Fig. 88? Describe the appearance of the carved ledge. Does blown sand produce similar effects in the eastern United States? See section 120. If possible visit an establishment in which the sand blast is employed. Learn the sources of the sand, the method of propelling it, and the purposes for which the blast is used.

CHAPTER VI

GLACIERS

121. **Intensive Study of the Gorner Glacier.**—Take first the map, Fig. 89. What is the scale, approximately? The contour interval? What is the color of the contour lines? For the rock? For snow and ice? For streams and ponds? What lofty peaks are named on the map? The Matterhorn is westward, off the field of this map, but in line with the main (Gorner) glacier (see Fig. 134). The dotted line is part of the Swiss-Italian boundary. Where is the Gorner Grat? Observe that this is the point from which the view in Fig. 90 was taken. It is about 10,000 feet in altitude, about 1,700 feet above the Gorner Glacier, and about 5,000 feet lower than the great summits from Monte Rosa to the Matterhorn. The student can thus appreciate it as an elevated position in the midst of a vast amphitheater formed by mountains and glaciers.

In what direction does the trunk glacier flow? What is it called at its lower or northwestern end? How many tributary glaciers are named on the map? Which is the most westerly? The most easterly? Which is the largest? Observe the spurs between the glaciers. What are the brown lines leading off from these? (See section 121.) How many of these? How many appear on the Boden Glacier? Why the diminution? How do you account for the abrupt ends of some streams on the glacier? Take Fig. 90. Can you tell where the snow-fields of the upper slopes merge into glacial streams as they descend? Why not? How many medial moraines are strongly shown? How many appear faintly? Why? Can you identify some of the strong medial moraines with those shown on the map? Some of the upper slopes which appear to hold snow support "hanging glaciers." This means that they rest on great ledges, and as they push down, masses of ice break from their lower edges and fall with the sound and appearance of snow avalanches. See the Matterhorn, Fig. 134. The hanging glaciers are well seen at the foot of the right-hand steep face of the mountain.

If the student has access to Tyndall's Glaciers of the Alps (edition 1896, Longmans) let him read the first ascent of Monte Rosa, 1858, pp. 122–133; the second ascent, most of the way without guide, pp. 151–160; and of the view from the Gorner Grat, pp. 137–140. Tyndall's three attempts upon the Matterhorn are found in his Hours of Exercise in the Alps, and the first ascent to the summit, by Whymper, is vividly told in Whymper's Scrambles in the Alps. In the summer of 1839 Louis Agassiz made a careful survey of the great group of glaciers lying between Monte Rosa and the Matterhorn. Zermatt, one of the chief tourist resorts of Switzerland, is in the Visp Valley about 1½ miles below the Boden Glacier.

122–124. **Study of Photographs of Other Glaciers.**—Specific questions can not be given, since such views as are available must be studied. The mountain slopes, tributary glaciers, moraines of various kinds, the surface of the ice and the stream issuing from the glacier should all be described, so far as these features appear.

125. **Greenland Ice-Sheet.**—The student will be able to answer the questions by reference to the following: section 125; the map Fig. 8 of this manual; and any one of the following: The International Geography, pp. 1040–1043; article Greenland in the New International

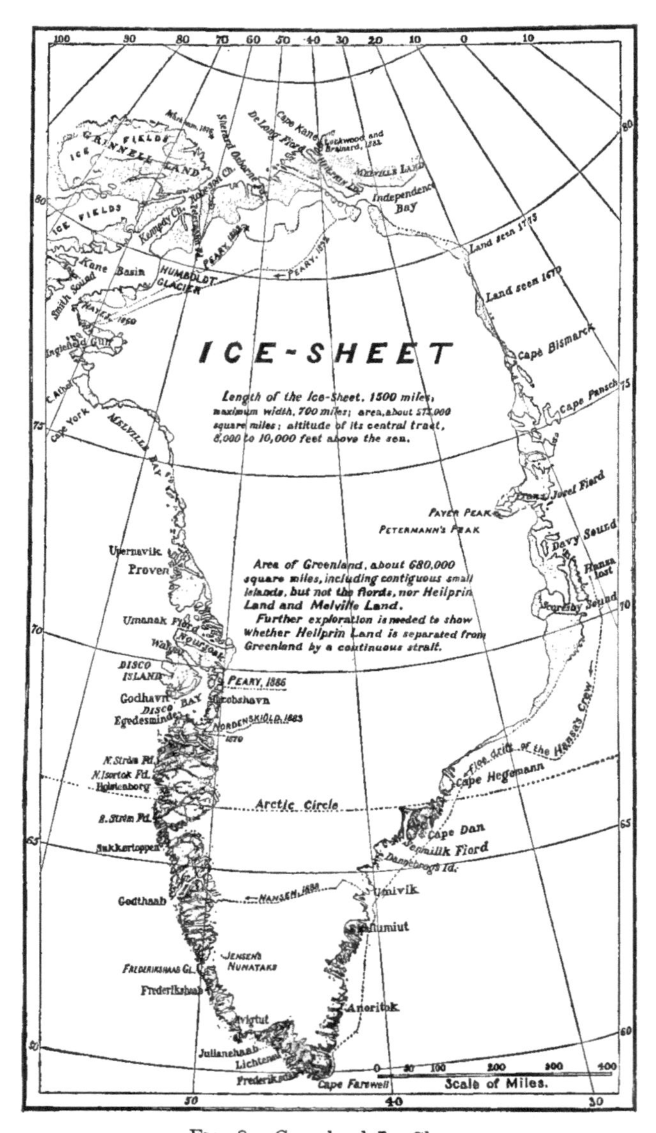

FIG. 8.—Greenland Ice-Sheet.

Encyclopedia; article Greenland in Encyclopedia Britannica, or other encyclopedias. Other references in Teacher's Guide, section 125. What are the length and breadth of Greenland? Note that answers vary somewhat with the authority consulted. What is the area? What share of the land is covered by the ice-cap? How wide are the ice-free borders of the island? Is this a rough or smooth shore-line? Consult the map. The student would find it useful also to read sections 259 and 263. What explorers have made journeys over the inland ice? Give routes, dates, and distances so far as they appear on this map. What is the character of the surface of the inland ice? How high is it known to rise above the sea level? Does it carry any moraines? Does the ice-sheet disappear entirely by melting on land? If not, how does the balance of the ice escape? What results from this? (Section 126.) How does this ice-sheet rank among those now existing. (Section 127.)

129. **Glacier Motion.**—Seek and record all proofs that you can find that glaciers flow. Consult the text-book, sections 121, 126, 129 and 133. For the use of instruments in determining motion and its rate, consult Tyndall's Forms of Water, International Scientific Series, pp. 59–72. This passage also gives proof from drifting of Swiss huts on the glacier. See also Brigham's Text-book of Geology, pp. 92–94.

Does ice appear more brittle than a stick of sealing-wax? Support the stick by its ends in a room of ordinary temperature and observe it at intervals for a day or two. Record the results. Test a stick of molasses candy by gentle pressure and by a sudden blow. A block of asphalt will serve the same purpose if a broad piece is rested on a narrow support for a few weeks and the adjustment noted. The student must not think that these illustrations *explain* the motion of glaciers. They only show that glacier ice is not the only brittle substance which will change form under pressure, gradually or steadily applied. When the ice of a glacier falls over a declivity (Fig. 250 and section 122) it breaks into great fragments. Below, it often closes up and may be readily crossed. The glacier is everywhere subject to powerful strains, and may be cracked or crushed at an infinite number of points, always freezing solid again. Compare the crushed ice of the cream-freezer, becoming a solid mass after a few hours.

132. **Till and Drift Boulders.**—If you find an exposure of stony clay, study it, in accordance with the following suggestions. Be sure to see whether it is a fresh exposure, or whether slip or washing has taken place. In this case you would not see the real conditions unless you dig into the bank. Do you observe any traces of stratification with sorting and layers? If so, how much and in what part of the bank? If there is none, record the fact. Examine the stones in the clay. Record their characters, whether many or few, large or small, angular or worn. Are any of them scratched? Are the flat stones on edge as much as in any other attitude? Compare this condition with the flat stones along the banks of a river or lake, or in a gravel bed. How many different kinds of stones can you find in the exposure? Do you know enough about the ledges and quarries within a few miles, or within 50 miles, to trace any of the pieces to their sources? Record the color of the deposit, or its colors, if there are differences. If your deposit is of clay, unstratified or very little stratified, and contains scratched stones, it is the "till" of the glacial geologists. Consult section 132 and observe Figs. 98, 99, 100.

Study the boulders in the fields and stone walls. Record the relative abundance of the more common kinds. Observe the resistance to weathering. Some will show little change, and others are ready to crumble or fall apart, from their exposure since the close of the glacial period. Select a few of the larger and measure their dimensions with rule or line. If a large boulder has been removed from its bed in the ground, see how much the upper parts with their long exposure have weathered in comparison with the parts protected by the earth.

133. **Field Study of Glacial Pebbles and Striated Bed Rocks.**—Gather as many as convenient of the scratched pebbles from the bed or bank of till. Select one which is somewhat longer in one direction than in others. In which way do the scratches run? Is the stone scratched on several sides? How could this be done? Do you find parts of the stone little scratched? If so, why? Are all the scratches parallel to each other? If not, what has happened? How much have the corners been rounded? Estimate what proportion of the bulk of the stone has been lost by glacial wear.

Compare this specimen with the others, and first in its form. To what extent are the scratches parallel to the long axis of the stone? Are some stones more rounded than others? Are some smoothed on more surfaces than others? Have some wabbled under the ice more than others? Note the kinds of rock. Do some take finer scratches than others? Are some kinds more numerous than others? If one or two kinds predominate in the particular bed, what does this show? Can you find any kinds which do not seem to have taken the scratches? If so, why? Do any of these seem to have been planed or shaped, without taking distinct scratches? If the collection of scratched pebbles is in the laboratory, they can be studied in a similar way, except that commonly they will not show the grouping or proportion of any one place.

Study a patch of scratched bed rock. This may be associated with the bank of till (see Fig. 100) or otherwise (Figs. 104, 105). What is the character of the rock? Would it scratch easily or with difficulty? Are the scratches fine and the general surface smooth, or are they deeper and forming small furrows? What is the direction of the scratches? Use the pocket-compass. Lay it in a flat place and let the needle come to rest.* Read the angle between the needle and the direction of the scratches. This may be gotten approximately by holding a straight edge, as a face of your lead-pencil, over the pivot of the needle and parallel to the scratches. In the United States it is convenient to read the angle with reference to the south, since the ice was moving in that general direction, unless shifted by local causes. Thus if the groovings have a direction of 18° west of south (or east of north) they would be recorded as S. 18° W. Note whether the rock surface is flat or on a slope, also any rounding of ledges or molding of hilltops (Figs. 104, 105). Would the scratches tell which of the two directions was taken by the ice? (Sometimes they would, but oftener not. Here call to mind what you have learned of the position of the bed rocks which have supplied the materials of the drift.)

134–141. **Field Study of Various Glacial Features.**—Here the material will be extremely variable, and only available, as in 132, 133, within the glacial belt. Definite instructions must be expected mainly from the teacher. The "washed" or water-rolled waste from the glacier is important. In every such case study and describe the external forms, whether kames (section 134), eskers (135), terraces, or deltas. Study also the materials and stratification wherever exposures occur. Gravel and sand-pits are prolific sources of knowledge. See if you can find any scratches on the pebbles and boulders. If not, why not? Can you find any faint scratches, showing partial obliteration by rolling in water. Study, as in 132,

---

* If the declination is large, take account of it. See section 237.

the kinds and proportions of the smaller and larger stones. Make diagrams of the strata, showing alternations of coarse and fine, and whether horizontal or inclined, with all the variations that appear in the section.

134. **Moraines.**—Eagle Sheet, Wisconsin. How many fairly distinct areas of marsh in the quadrangle? How many square miles of marsh by estimate? What share of the entire quadrangle? How many lakes and ponds are shown on the map? Note any of them that may have been much larger formerly. What has reduced their size? What feature of the contour lines attracts attention? Do the elevations show system or otherwise in their arrangement? Give attention to that part of the map to the northwest of the Chicago, Milwaukee and St. Paul Railroad. How do "depression contours" differ from others? See "legend" on right margin of sheet. How many of these depressions are shown on the part of the map designated? Do any of them contain ponds? How many? Why not the rest? Can you draw a conclusion as to whether the deposits are of sand and gravel or otherwise? Returning to the map as a whole, can you find a lake of some size which has no surface outlet? What kind of contours surround it? Why does not the basin fill up? How high do the hills range above Fox River? How deep are the deeper depressions in the hills, as shown by the depression contours? These depressions, or "kettles," have led to the common use of the name "kettle moraine" for the great belt of such heaps of waste in the State of Wisconsin.

134*a*. Charleston Sheet, Rhode Island, and Stonington Sheet, Rhode Island, Connecticut, and New York. Where on the Charleston Sheet do the contours suggest a belt of moraines? About how wide is the belt from north to south? How high do the hills range above sea level? Observe Watchong Pond. What brook and river drain it? Trace thus the drainage of Pasquset and Worden's Ponds. What swamps along the range of these ponds? Note Wood River, Beaver River, and Chipuxet River. What carries their waters to the sea? Follow this main

FIG. 9.—Esker near Freeville, New York. Height thirty to fifty feet. Photograph by Prof. Frank Carney.

stream as shown on the Stonington Sheet. How does the region south of Westerly compare with the moraine belt of the Charleston quadrangle? Why is so much of the drainage diverted to the westward? Describe the contours on Fisher's Island. How does Fisher's Island compare in direction with the mainland east of Watch Hill?

135. **Eskers.**—Study Fig. 9 of this manual. Where is the esker? Trace in your note-book a curved line, which, as nearly as you can judge, represents the changes of direction of this esker. In doing this, note the swing from right to left at the extreme right of the view. What angle do the side slopes make with a horizontal plane? Roads are often carried along such ridges to avoid marshy ground on either hand. Would this crest be wider than would be necessary for such a purpose? What is the height of this esker? (See title.)

If access can be had to an esker in the field, the study can be carried further. Note all the points called for above. Plot the course of the ridge more in detail than is possible from a photograph. (The esker of the figure extends considerably beyond the farthest point in view.) Ascertain in detail the fluctuations in height. Seek for exposures of the material, and study its character and structure, as in exercise 134–141. Has the esker determined any human feature, as roads, lines of fence, buildings, etc.? What origin for such ridges is suggested in section 135? Could a stream flowing under ordinary conditions make such a deposit? If confined in a tunnel, or in an ice canyon, what change would take place in the deposit when the ice-walls melted away?

136. **Map Study of Drumlins.**—Sun Prairie Sheet, Wisconsin. Locate the sheet in the general map. Estimate the ratio sustained by the marsh lands to the whole. Compare as to interruption of drainage this quadrangle with that shown on Eagle Sheet (exercise 134). Study the drumlins. What is the general direction of their longer axes? What is the relative length and width of the average drumlin shown here? What is their average height above the swamps, or above the dry lowlands from which they rise? Can you find any that are about 250 feet in height? Can you detect any system in the pattern

of the drainage? Would you expect any regularity? How would the streams compare in this respect with those of Eagle quadrangle? What is the relative frequency of lakes in the two areas? Can you find any depression contours on this sheet? Have the drumlins influenced the course of the railways to any degree? Where the roads have not swerved, are the necessary railway cuts shown by the contours? How many apparently well-formed drumlins in Deerfield Township? How many square miles in this township? What does the shape of the drumlins indicate as to the agency of the glacier in making them? Compare the rock dome in Fig. 105, remembering that the latter is high and steep-sided for its kind. The regular curvature of the contours shows very smooth surfaces, not gashed by small streams. What testimony does this offer as to the age of the drumlins? If the adjoining sheets are at hand, they may be studied in like manner. They are: Waterloo, Watertown, Evansville, Stoughton, Koshkonong, and Whitewater, all in Wisconsin. Compare the trend of the drumlins of the Madison Sheet on the west, and then observe the shape and trend of these hills on the Watertown Sheet, the second to the east. On the last sheet study the course of the Rock River, east and southeast of Watertown.

136*a*. **Drumlins of Western New York.**—Many sheets are named in the Teacher's Guide. These may be grouped for general study, or one or two of the best may be studied. The method has been sufficiently shown in 136. Take, for example, the Oswego Sheet and note trend, height, proportionate length and width, and the cutting away of parts of three large drumlins by wave action northeast of Fairhaven. The Baldwinsville Sheet to the southeast shows a slightly more eastward trend and the less symmetrical or less developed form of many of the hills. Clyde Sheet shows the drumlins of the shore region farther west. Note the change in the form of the drumlins toward the sound end of the area. Connect with this the Geneva Sheet. Where do the drumlins disappear? Are there any well-formed drumlins in the Geneva quadrangle? The Auburn quadrangle shows similar features.

If drumlins are near the school, let them be visited and studied in detail. Make longitudinal and cross profiles, and study the materials if exposed.

141. **Glacial Lakes.**—Plymouth Sheet, Massachusetts. Describe this area as regards its general surfaces, its altitudes, and the forms of the hills. What are the dimensions of the longest lakes? Of the smallest ponds? What share of the lakes and ponds have surface outlets? Observe the small quadrangle enclosed by the meridians 70° 35′ and 70° 40′ and by the parallels 41° 45′ and 41° 50′. How many square miles? How many bodies of water? How do the roads show the character of the region? Compare the stream courses with those of Sun Prairie and Eagle quadrangles, Wisconsin. What other sheets already studied show a considerable number of glacial lakes? Connect with this the Falmouth Sheet, if at hand, and make similar observations.

If you have Paradox Lake Sheet, New York, compare it with Plymouth Sheet. What of the relative size and number of the lakes? Are there so many which have no outlets? Describe the elevations among which the lakes lie. Observe that here we have lakes in mountain valleys, due to the clogging of these valleys in places with drift.

CHAPTER VII

PLAINS

150. **The Atlantic Coastal Plain.**—Review carefully section 52 and Fig. 47, observing in the latter the extent of the plain to the fall line. Review section 59. Read in advance section 262. Study Fig. 10 of this manual. This does not show any particular locality, but represents any coastal plain in its general character and relations. Note the present position of the shore-line. Explain such indications as you see that the sea has swept farther in upon the lands. What share of the area shown in the model is now sea? What share was formerly sea? What has been the cause of the change? Is there any evidence that the sea-border kept its inner position for a considerable time? Were the rivers then at work in the mountain region? Where was the land waste of that time deposited? What change in the rivers with the uplift of the land? What changes are taking place in the new lands? Would the materials of the coastal plains wear more or less easily than those of the uplands? Why are the valleys so shallow? Why are the branches of the main stream so few and short? Why is one valley of the plain wider than the others? Does the tide enter these rivers?

150*a*. Wicomico Sheet, Maryland. Locate on the general map, and record scale and contour interval. What are the two chief rivers? Are they tidal or otherwise? What are the areas shown in blue with fine horizontal lines? See legend. Both in this map and in the model are represented lands which formerly were sea bottom, now uplifted. Why are the rivers tidal, with salt marshes, in

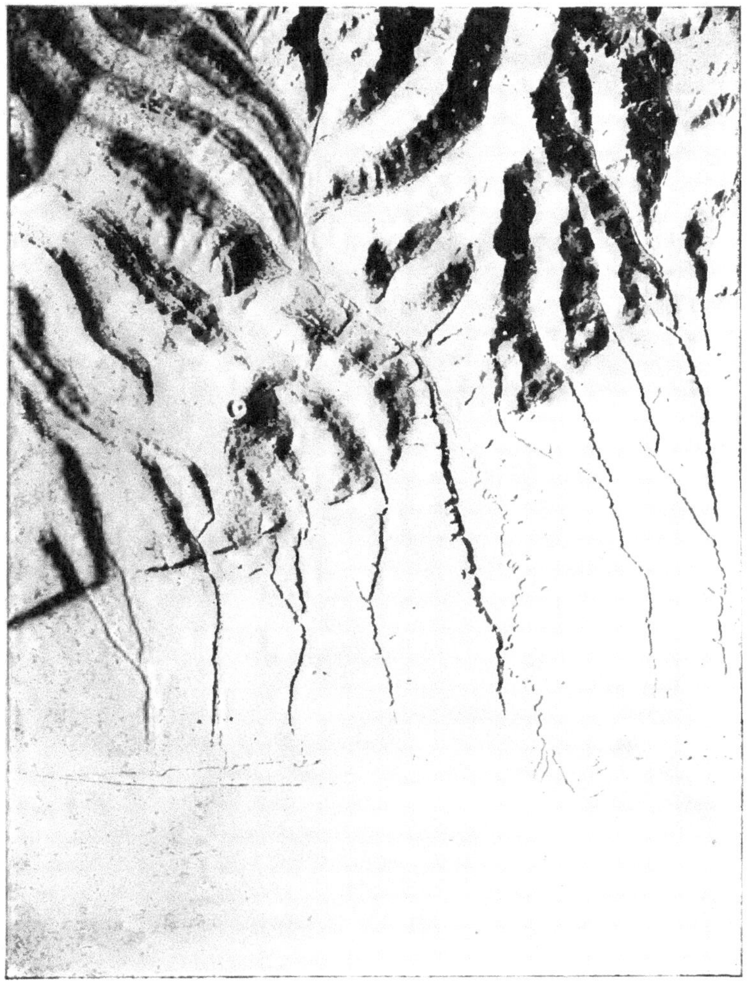

FIG. 10.—Harvard Geographical Model, No. 2.

this case, and not in the other? What kind of ground lies upstream from the salt marsh along Chaptico Creek? Why the difference? At what other places on the map are the same conditions seen? What is the altitude of the highest lands in the quadrangle? What is the character of these surfaces? Have the streams any considerable flood-plains? Are the sides of their valleys steep or otherwise? Are the branches of the main stream short or long? What is the meaning of these facts? Are the upland plots changing in form or size? How are these plots drained? Where do the roadways run? Why? Observe the highways leading out in every direction from Chaptico. Do these roads afford any exception to the general rule? Can you imagine a time when it might be necessary to shift the roads to the valleys? Why? Note the courses followed by the highway and the railway northward from Pope Creek.

The student may refer to Fig. 44, which shows an area a short distance down the Potomac on its right bank. Compare exercise 45–50. It would be well also to read A Coast Swamp, Topographic Atlas of the United States, Physiographic Types, folio ii. We find here the Norfolk Sheet, with the adjoining parts of the coastal plain and a part of the Dismal Swamp.

152–155. **Map Study of Lake Plains.**—Niagara Sheet (studied for falls and gorge in 32), Wilson and Tonawanda Sheets. Read in advance section 144 and observe Fig. 112, noting particularly the position of the "Ridge Road." Find the Ridge Road on the Tonawanda Sheet (Dickersonville and Cambria) and trace it westward across the border of the Niagara Sheet to Lewiston. Note that Lake Iroquois, the greater Lake Ontario of former times, shown in the map, Fig. 112, occupied all the space between the Ridge Road and the present lake shore. How wide is the belt? What deposits would it receive when the lake water covered it? Would it thus become smoother or rougher land? How is it now? How much descent is there from the Ridge Road to the lake? How much is this per mile? What kind of a grade has the Rome, Watertown and Ogdensburg Railway? How does the area compare in the pattern of its roadways with the Eagle or Sun Prairie quadrangles? Are there many streams? Have they to any degree excavated valleys? What do these conditions mean?

Review the Fargo Sheet. It was studied (exercise 13) as a sample contoured map, but it was also explained that the region is young, and its drainage imperfectly developed. In this respect it may now be associated with the Iroquois plain and the Atlantic coastal plain. Read sections 152, 153.

152–155*a*. Tooele Valley Sheet, Utah. Read section 154 and observe Figs. 116, 117. What lake appears in part here? What are the scale and contour interval? What is the direction of the mountain ranges? How high are they? How wide are the lowland belts between them? In an ordinary mountain country would you expect to find the lowlands so even? What has happened here to obliterate the inequalities? Great Salt Lake is but about 40 feet deep. If it should dry away would a basin appear, or would the bottom be continuous with the Lake Bonneville bottoms as we here see them? Is there any other reason than youth for the fewness and smallness of the streams in this region? Remember Great Salt Lake Valley, Red River Valley, and Lake Ontario lowlands as three typical lake plains (Lake Bonneville, Lake Agassiz, and Lake Iroquois).

159. **Prairie Plains.**—Dunlap Sheet, Illinois, and Fig. 11. Note that Dunlap Sheet and La Salle Sheet (from which Fig. 11 is taken) are both on the Illinois River. Record their position in relation to each other, also in re-

lation to Chicago and the month of the Illinois River. What city is situated a short distance south of the Dunlap quadrangle? What is the scale on this sheet? What is the contour interval? Why is this desirable? What range of altitudes do we find in this area? Where is the lowest point? What is the general altitude of the upland areas along the divides? Are the divides sharp or ill-defined? How many streams have cut broad valleys? How high are the bluffs west of the Illinois River? How far west of this bluff is the prairie considerably trenched or dissected? Have the streams that have done this work developed flood-plains? Will they do so? Will they become longer? What evidence of a change of course by the Illinois River? What do the depression contours in this belt mean? How high does the ground between the old channel and the new rise above the old river bed? Has the river deepened its channel since the change, and if so, how much? Is the river sluggish or swift, and what is the evidence? What streams west of the bluffs have developed flood-plains? How do this topography and drainage compare with those of Fig. 41? Is this an old or a young land surface? Is it less or more mature than the Fargo quadrangle?

What is the general arrangement of the roads in this area? Compare in this respect with Fargo, Sun Prairie, Eagle, Plymouth, and other quadrangles. Why is there departure from this arrangement along the Illinois River bluffs? Which prevails on the river lowlands? Why change from east and west to the east of Kickapoo? Is the region thickly settled or otherwise? How many families on the average on a section (square mile, see scale)? What relation between this and the character of surface and soil? The soil and underlying drift in this region are so deep that wells about Dunlap have been sunk from 100 to 150 feet in depth without reaching bed rock.

Fig. 11, or the entire La Salle Sheet, if at hand, may be studied. Record depth and width of main valley, short narrow gorges and broad smooth inter-stream uplands, etc. Students living in Illinois will find it of interest to consult The Illinois Glacial Lobe, by Frank Leverett, monograph xxxviii, United States Geological Survey. It contains in maps and text a great body of facts about the rocks, drift, soils, streams, and wells of the State.

CHAPTER VIII

MOUNTAINS AND PLATEAUS

163–164. **Profile of Rocky Mountains in Colorado** (Fig. 123).—Use horizontal scale, same as in figure; vertical scale $\frac{1}{8}$ inch = 5,000 feet. Represent sea level by a base-line 4 inches long. Make first profile east and west by Colorado Springs. Locate along the base-line Colorado Springs at the east base of the mountains, Pike's Peak summit, middle of South Park, crest of Mosquito Range, Arkansas River, crest of Sawatch Range, and two or three minor crests along the line to the westward. Colorado Springs is 6,000 feet above the sea, and the plain slopes eastward gently. Pike's Peak has an altitude of about 14,000; South Park and Arkansas Valley, 8,000 to 9,000; the Sawatch Range about 14,000, and Mosquito Range about 11,000. The vertical can be erected in accordance with these figures and the profile drawn.

Make a similar profile from east to west along the line of the word "park" in San Luis Park and compare it with the first, noting differences. Compare your profiles with the geological section of the Front Range (Fig. 125). Where does this section belong, if its line were drawn on the map (Fig. 123)? In learning the structure of these mountain ranges from the text and from Fig. 125, consult also Gannett's Physiographic Types, folio ii, Hogbacks. His map, view, section, and text explain quite fully "Rocky Mountain structure."

165. **Colorado Plateaus.**—Echo Cliffs Sheet and Kai-

bab Sheet. If the group of sheets noted in the Teacher's Guide is available, the study can be made to cover the entire region. Review exercise 27*a*. Note scale and contour interval. What is the altitude of Paria Plateau? The direction of slope? Height of Vermilion Cliffs? For a general view of such cliffs see Fig. 126. Altitude of plateau south of Vermilion Cliffs? Depth of Marble Canyon? How wide is the dissected belt along the Marble Canyon? Along the Grand Canyon? Why this difference? What considerable streams enter the Colorado within

FIG. 11.—Block Mountain Structure.

these two quadrangles? Do they enter the trunk river at grade? Is there any exception—that is, does any stream show rapids or falls near the junction? Does any perennial stream enter the Colorado between Paria Canyon and Little Colorado River? How far is this? Between Navajo Creek and Little Colorado River, as shown on Echo Cliffs Sheet, are about 600 square miles of territory without a perennial stream. Compare the streams on an area 20 × 30 miles in size near your home.

Is this a mature or an immature land surface? What is the climate here? How would this affect the rate of maturing? This surface is much older than that of the prairies in Illinois owing to the drift cover in the latter regions. Why has the Colorado no apparent flood-plains, while those of the Illinois River are broad? What bearing has material on which the rivers work upon this question? What importance has the difference in altitude in this question? The size of the two streams? Why are the springs all marked on the two sheets?

166. **Great Basin Topography.**—Fig. 11 of this manual. Figs. 127, 128, 129. Disaster Sheet, Nevada. Con-

sult also Tooele Valley Sheet, Utah (exercise 152–155*a*), and Granite Range Sheet, Nevada.

In Fig. 11 observe that the rocks are represented as stratified. How much have they been inclined from their original position? Along how many planes have they been fractured in the making of the uplift? Such a dislocation is called a "fault." If the fault is large in amount—some hundreds or thousands of feet—mountain ridges may result. Such are sometimes called block mountains, because consisting of blocks of elevated and tilted strata. Suppose we have cross-sections and profiles of such mountains in our figure? Are the slopes equal? Estimate the angles of slope on the two sides. Where will the waste from weathering accumulate? In the Great Basin would any of it go to the sea? Does the figure show accumulations of land waste? How many faults are shown in Fig. 127? How much are the strata tilted as compared with those of the previous diagram? Observe the bed numbered 4 in each block. In which block is 4 the highest? What has happened to the strata above 4 in block B? What strata have been removed from block C and block D? Is the land waste shown in this diagram? Which side of a tilted block have we in Fig. 128? Is there any waste at the foot of the cliff? Would you call this an old or a young block mountain? Observe Fig. 129, but only such parts as lie between the Great Salt Lake and the Sierra Nevada. Here are the Basin Ranges. What is their general direction? They mark therefore the direction of the great faults by which the mountains were produced. The flat belts between are areas of waste, in some places accumulated in parts of a large, straggling lake, existing here when Lake Bonneville occupied the Utah Valley.

Study the Disaster Sheet. How many mountain ridges appear in the north part of the sheet? Compare the slopes on the opposite sides of the two easterly ranges along the parallel 41° 45′. Reckon the amount of slope per mile in each case. Make a profile across the map along this parallel. Use the horizontal scale of the map, take sea level as a base-line, and select a suitable vertical scale. What indication does the drainage furnish of the climate of the region? Can you suggest any connection between the existence of hot springs here and the mode of origin of the mountains? Has the cliff north of Thacher Pass suffered much dissection? What conclusions from this?

167. **Physiography of California** (Fig. 129).—Observe the California-Nevada boundary-line. How many of the Basin Ranges cross this line into southern California? What is the direction of the Sierra Nevada Range? Are its slopes equal? Which is the steep side? If the relief were shown by contours, how would the arrangement of lines differ on the two sides? What valley lies west of the Sierras? How is it drained? See section 61 or any atlas for names of streams. What range lies between this valley and the sea? How does it compare in height with the Sierra Nevada? Does this map show that the Sierras are higher? From which range do the rivers of the valley receive most of their water? How are the flat lands of the valley formed? See sections 79 and 156. What break in the Coast Range? This is like the fiords of the New England coast. What is thus suggested as to this coast region? Is there a coastal plain between the Coast Range and the sea? Would there be one if there should be an upward movement of the lands?

In this connection refer to Gannett's Physiographic Types, folio ii, Alluvial Cones. Observe the locality south of the San Bernardino Mountains and east of Los Angeles. It is thus outside the great valley, but shows the same features. These are—mountains whose torrents form broad, low alluvial plains; an arid climate, making the stream waters valuable for irrigation; and dense population; while the neighboring mountains bear great forests, now to a large degree held as forest reservations.

Adirondack Mountains.—Mount Marcy Sheet, New York. Describe the relief of the region. What is the altitude of the main valleys, as at Keene Valley, Ausable Lakes, and South Meadow? Height of Mount Marcy? Of other summits? Study the drainage in connection with a map of New York, and show where these headwaters belong. Do the mountains and the drainage show any system, and if so, what? Make a tracing of the drainage, and sketch in the lines of water-parting. Estimate the average altitude of the quadrangle above the sea. Compare the altitudes of valleys and summits with those of Colorado. What evidence that some of the lakes have been partly filled? Have any lakes been completely filled? What tends to preserve the slopes, and, for the greater part, the tops of these mountains, from wasting? Make a profile across the summit of Mount Marcy and the head of upper Ausable Lake, using horizontal scale of the map and vertical scale slightly exaggerated, $\frac{1}{8}$ inch = 500 feet. Where are the principal permanent settlements? What must the human interests of such a region be? What interest is prominent in the Rocky Mountains and nearly absent from the Adirondacks? Find North Elba. Some students will be interested to know that this is the burial-place of John Brown.

170. **Appalachian Folds.**—Harrisburg Sheet, Pennsylvania. Review Pine Grove Sheet, exercise 51. Before studying the Harrisburg area turn to Fig. 131. This is from the neighboring State of Maryland, but represents the continuation southward of the same kind of folding that has taken place about Harrisburg. The different beds are not numbered, but are shown by different signs. How

many up-folds are shown? How many down-folds? At what angle of inclination do you find the strata that have been tilted most? The student should remember that beds are often tilted more, even to the vertical, or overthrown, and that all these attitudes may be found in a single mountain range. How many remnants of the upper strata are found? How many of the second and third from the top? How much does Warrior Ridge rise above sea-level? What was its original altitude according to the diagram? What sort of beds would survive wasting and stand out in relief?

What is the character of the surface between Harrisburg and Blue Mountain? What is the height of Blue Mountain east and west of the Susquehanna River? What is the average rise of its slopes in a half mile? What are the other mountain ridges of the area? What is their relation to each other? How far apart are their crests? Observe Stone Glen, about 3 miles from the Susquehanna on Stony Creek. Make a profile across the four ridges, using horizontal scale of the map and vertical scale, $\frac{1}{16}$ inch = 100 feet. How much vertical exaggeration does this involve? The ridges are determined by resistant outcrops of hard, thick beds of Medina sandstone.

Make a tracing of the drainage. Review exercise 51. This is an old region. Why are not the branches of the longitudinal streams, Fishing Creek, Stony Creek, etc., more developed? On such short, steep slopes could any single tributary gather much water? What water-gaps have we here? Consult section 51 and Fig. 45. Note that the Delaware Water-Gap is cut through this same ridge far to the northeast. How wide is the gap at Fort Hunter? How deep? Make profile across it along the crest of the ridge.

On a general map look for other water-gaps between the Susquehanna and the Delaware. Harrisburg lies in the "Appalachian Valley." Is the region about it thickly settled? What other important towns lie in this valley in Pennsylvania, Maryland, and Virginia? (In the last

Fig. 12.—Kaaterskill Clove.

the valley has the name Shenandoah.) The rocks of this valley are limestones and shales. What influence would this fact have on the soils? What minerals are found in this part of Pennsylvania, including the mountain-belt? To what great city is the region tributary? What influence have geographic features had upon the routes of travel and commerce in this region?

173*a*. **Allegheny Plateau.**—Catskill and Kaaterskill Sheets, New York. Fig. 9 and Fig. 12 of this manual. Note scale in the sheets and in Fig. 9. Observe in both the Catskill escarpment or great cliff facing the river. Which gives the most striking impression of the topography? Which gives the most accurate knowledge? How far is the escarpment from the Hudson River? How high is it? What is the character of the uplands on the west? Does a plateau necessarily have a smooth surface? What is the origin of Kaaterskill and Plaaterskill Cloves? Compare these as seen on the map in Fig. 9 and in Fig. 12 of this manual. The last is Kaaterskill Clove, often named from the village of Palenville. Is the clove still growing in length? See Fig. 12. What is the position of the strata? Observe the head-waters of Schoharie Creek, and trace their course to the sea on a map of New York. What is the character of the lowlands between the plateau and the river? Note in Fig. 9 and on the map a succession of ridges, or a broken ridge, about 3 miles from the river. These are sometimes called the Little Mountains. How high are they? They show true Appalachian or folded structure and are real mountains, although so much lower than the Catskills, which, as we learn now, are not true mountains. Make a profile showing plateau, escarpment, lowland, and river, selecting a suitable vertical scale.

173*b*. **Cumberland Plateau.**—Chattanooga Sheet, Tennessee. Suwannee, Cleveland, and Ringgold Sheets on the west, east, and south would be useful, also McMinnville Sheet. Use the Chattanooga Sheet. What are the scale and contour interval? What is the altitude of the valley along the Tennessee River? Study the escarpment at the foot of which runs the Cincinnati Southern Railroad. How high is it? What is the altitude of the plateau to the west? To what extent is the cliff broken by gorges?

We have here the Cumberland plateau and the Cumberland escarpment, corresponding to the Allegheny plateau and the Catskill "Mountain" front in the north. What break in the plateau to the west? Is the plateau resumed beyond this valley? What name is given to the narrow plateau belt between the Tennessee and the Sequatchie? Observe that the Tennessee leaves the Appalachian Valley at Chattanooga and enters a gorge in the plateau.

Consult Fig. 133. What elevation is south of Chattanooga? How does this change going southward? Observe a valley west of it. It is the valley of Lookout Creek. Lookout Mountain is a spur of the great plateau of which Walden Ridge is also a part. In the plateau the rock-beds are horizontal. In the valley they are upturned. Compare Catskill plateau and the Little Mountains to the eastward. If you have the Cleveland Sheet, note the eastern border of the valley along Beans Mountain. How is the Tennessee chiefly crossed? Some of these ferries, as Brown's and Kelly's west of Chattanooga, were of great importance to the Federal army in the civil war. Lookout Mountain, Missionary Ridge, and Chickamauga are also famous in this connection. The great battle of Chickamauga was not fought near the railway station of that name, but farther south and west. On the Ringgold Sheet the battle-field can be located a short distance southeast of Rossville Gap. Are individual houses located on this map? Would the scale make it practicable? How are the towns located with reference to lowlands and uplands? How are the towns along the Cincinnati Southern Railroad located with reference to streams? Can you suggest a reason in this case? What reasons for a city on the site of Chattanooga? What other important town in the great valley in the same State? What great southern city to the southeast? In what directions from Chattanooga do geographic features admit of free communication? The McMinnville Sheet can be studied as a typical part of the plateau. Record altitudes, behavior of larger streams, the finger-like or branching drainage at the northwest, etc.

173*c*. **The Plateau in West Virginia.**—We have studied it in the north and in the south, and we now take an intermediate locality. We have had a sample of it in Fig. 41 and in exercise 45–50. Examine the following sheets: Charleston, Kanawha Falls, Nicholas, Oceana, Raleigh, and Hinton. Record your observations on the altitude of the plateau, the depth of the Kanawha Valley and the chief tributary valleys, and on the completeness of dissection. Would table-land serve for a synonym of plateau in this case? How are the railways and towns located?

Taking the entire plateau, from north to south, what States belong in part to it? How does it vary in form of surface? What mountains lie to the east of it? (Observe that the great valley which we have seen at the Catskills, at Harrisburg, and at Chattanooga is a part of the mountain country much reduced in level.) What great lowlands lie to the northwest of the plateau?

CHAPTER IX

VOLCANOES

185. **Mount Shasta.**—Mount Shasta Sheet, California (Fig. 144). Have at hand also Diller's Mount Shasta, No. 8, National Geographic Monographs, and Gannett's Physiographic Types, folio i, A Young Volcanic Mountain. Review the study of the Shasta Sheet in exercise 13. There it was considered mainly as a sample contoured map. Here we observe chiefly the features of the volcano. Read section 185. Why is Shasta a young volcanic mountain? Has dissection or denudation of the sides and summit made much progress? Are there lava streams which appear comparatively fresh and recent? See Diller, pages 245–250 and Fig. 5. Are any small cones preserved on the sides of the mountain? See map and Diller, pages 250–253. Give the substance of Diller's notice of lava caverns, pages 254, 255. What zones of climate does Diller describe, pages 255–257? What effects upon topography and vegetation follow? See also pages 265–267. Give the number and size of the glaciers. Record what you can learn of the glaciers from the map alone, and then add any important facts from Diller. Why are the glaciers distributed as they are? How does this affect the distribution of surface streams? Account for subterranean drainage. Turn to Fig. 148. Which stage is more nearly represented by Shasta?

189. **Lava Sheets.**—Belchertown, Northampton, and Springfield Sheets, Massachusetts. Fig. 13 of this man-

ual. Record the altitude of the Connecticut Valley low-lands at Springfield, Holyoke, Northampton, and South Hadley. Study the Holyoke Range. In what direction does its southern portion extend? Its northern portion? How long is it as seen on this map? What are its highest points and its average height? What streams pass through it? Are there other low gaps? What use is made of these gaps? Which is the steepest side of the range at Mount Norwattock? At Mount Holyoke? At

FIG. 13.—Faulted sediments and lava sheets of the Connecticut Valley.

Mount Tom? Northward from the Westfield River? The student will now remember that a very steep slope or cliff can not be adequately shown by contours, and may note that these western slopes are often cliffs that approach the vertical, so that the real contrast between the opposite sides of the ridge is much greater than appears. Consult Fig. 13 of this manual. What does the heavy black belt represent? Observe the dislocation or faults. To what angle have the fault blocks been tilted? How much denudation has there been? What rock forms the mountain ridges or hills? The student should now show how the diagram represents a succession of such ridges as we have upon the maps. Where, on the map, is shown the outcropping edge of the Holyoke lava sheet? Where is the "dip" slope? What has become of the shales and sandstones that lay east and west (or above and below)?

189*a*. **Lava Sheets.**—Tarrytown Sheet, New York; also New York and vicinity. Special map, combined sheets (Figs. 150, 151). Take first Fig. 150. You are looking up the Hudson, with the Palisades on the west. What is the origin of the talus? What share of the cliff does it hide? What is the appearance or structure of the rocks in the cliffs? Observe the columnar effect in Fig. 151. Similar structure is found in the bluffs on the west side of the Holyoke Range. On the special map of New York and vicinity observe the site of Yonkers. How does the land slope here? What conditions do you find across the river? Is the bluff shown only by contours? What part of it is shown in another way? What is this symbol called? See section 13. Why is it used here? Do the hachures show you the height of the Palisades? Can you find this out from this map? How? How far south can you trace this cliff? How far north? See Tarrytown Sheet. Does it change direction at the north end? Is there any feature like this in the Holyoke Range? How does the contrast between opposite slopes here compare with that of the Holyoke Range? What kind of ground lies west of the Palisades Ridge in New Jersey? What marked gap in the ridge? What advantage is taken of this? How do the railways from the west reach the river side of the ridge? In which direction is the Palisades cliff retreating? The Holyoke cliffs? What becomes of the Palisades lava west of the eastern border of the marshes in New Jersey? The student may note ridges of similar origin in the Watchung Mountains near Orange, Montclair, and Paterson.

190. **Crater Lake.**—Crater Lake, special map, Oregon, or, A Crater, in Gannett's Physiographic Types. folio ii. How deep? What is the altitude of its surface? Has it an outlet? How much larger is its drainage basin than the surface of the lake itself? Describe the land surface adjacent to the lake. How high are the cliffs? Describe the slopes leading off from Mount Mazama. What are the height, form, and position of Wizard Island? See map

and Fig. 10 in Gannett. What are the rocks about Crater Lake? See section 190 and Gannett. If you carry the slopes of Mount Mazama inward and upward what would you have? See Fig. 152 and compare Fig. 153. Make a profile similar to that at the bottom of the map, and then by a dotted line complete the outline of the ancient volcanic peak. If this were completed, what other volcanic peaks of Oregon and Washington would it resemble? Observe the gorges south of the lake, as seen in Fig. 152 and on the map. Where do they head? Could they have been cut out like this with the lake basin as it is? The slopes outside of the rim are glaciated, but not the cliffs about the lake. Could this condition have come about with the mountain as it is? What is the accepted theory of the origin of this lake basin? See section 190 and Gannett.

CHAPTERS X AND XI

THE ATMOSPHERE

Observation of Weather Conditions without Instruments.—The directions given to students must here be somewhat general, dependent on the teacher's plan of work. If the course in geography runs through the school year, the chapters on the atmosphere may be reached in February or March. But simple observations like those of this section may well be assigned early in the year, as for a single month, perhaps October, in the autumn. The student's record may for convenience be made at some stated hour of the school period in each day. It would then reach back to the same hour of the day before. A better plan, if the necessary attention can be given, is that the student's record be made at home in the early evening, after sundown, covering the weather conditions of the same day only. If this work be supplemented by a few moments of questioning in the class once or twice a week much progress will be made before the more systematic study, with instruments, is begun.

The body may be regarded as a kind of thermometer (section 207), but it gives us no figures. We know the temperature in a general way by our feelings. The questions that follow relate, of course, to out-of-door conditions. At the time of your record, does the air seem to you hot or cold? (If after sundown, it would rarely be the former.) If neither, is it warm or cool? Are you chilly or just comfortable in your ordinary clothing?

Have you made any change in the weight, or, as you say, the warmth of your clothing during the day? *How* does added clothing keep you from being cold? How did the air feel, as regards warmth, after breakfast? At noon? At two o'clock? What conditions of the air would affect your sensations of heat and cold? What would be your sensations if the wind were blowing on a warm day? On a cold day? Would the same effect be produced if you were riding in an open carriage? Why would a strong wind be more chilling than a breeze? Is there any other condition of the atmosphere besides its actual temperature that affects our feelings of heat and cold? What is it in the atmosphere that makes us call a hot day "sultry"? Why do we say some cold days are unusually "chilly" or "frosty"? The technical word here is "humidity," and fuller study will be given in sections 199, 200. A resident of Roanoke, Va., recently told the writer that he had suffered more keenly there at a temperature of 10° above zero than he had from very low temperatures when living in Iowa. What could make the difference?

Construct a table in your note-book for a record of temperature. It may be as follows:

DATE.	Morning.	Midday.	Evening.	REMARKS.
Oct. 1	Cool	Warm	Chilly	Began to be cloudy about 3 P.M.
Oct. 2	Cold	Cool	Cold	Strong north wind all the afternoon. Clear sky.
Oct. 3				
Oct. 4				
etc.				

The above is suggested for a single record to be made from memory at evening. If the student will undertake more precise work the record will be improved by substituting 7 or 8 A. M. and 2 and 8 or 9 P. M. for the more vague "morning," etc. But this is really important only for instrumental observations. As we already see, temperature is only one of the weather conditions of which our bodies give us knowledge. In our brief record above, for example, winds and cloudiness are noticed. Let the student take up these further conditions in a simple manner. Have there been winds? Were they light or strong? warm or cold? and what was the direction from which they came? Was there rain, snow, or hail during the day or during the previous night? How long did the rain or snow continue, and did much or little fall? Was any part of the day cloudy, and if so, were the clouds dense and appearing everywhere or was the day only partly cloudy? Record the more important of these facts under "remarks." You are thus to give in your table such a short, simple history of the day's weather as you can put down in five or ten minutes, as the weather has seemed to you, without reference to a thermometer or any other apparatus.

The temperature, the winds, the clouds, and other elements which make up weather and climate will be given more full and exact study as we come to them in later sections. But the habit of seeing and remembering can be fostered here. Under the head of remarks, any one day may be compared with some previous day of the record, as to its temperature, its rainfall, or the condition of the sky. And it would be well at the end of the month to give a summary or general account of the weather during the month.

It would be useful for the student to accompany such a series of observations by the preparation of a simple essay upon man's dependence upon weather conditions and weather changes. Not much time should be given to it, but an hour may be set aside in which the student may recall and put in order the simple facts already known to him. The following questions will aid in doing this:

What inconvenience would the farmer in the Northern States suffer if rains and cold weather prevailed until the 1st of May? Why would this be preferable to an "early" spring followed by frosts in May or June? Why would prolonged drought in May and June be more disastrous than the same conditions in midsummer? In what part of the summer would continued rains be harmful? What weather conditions are hostile to traffic on rural highways? In what ways is railway transportation endangered by prolonged rains? Are floods more dangerous to railways in a warm or in a cold climate or season? What losses are likely to result from a delay of one or more days in the forwarding of railway trains? What weather conditions are least favorable to lake and sea traffic? Give reasons why continuous rains would embarrass the movement of armies. Give your opinion of various weather conditions upon health. (This is a somewhat technical and difficult theme. It is only expected that the student's attention will be fixed and that such impressions as he may have will become the occasion of discussions. Thus he will have his impression of the effect on health of a severe winter, of the changeable weather of a Northern "open" winter, of a dry climate, etc.)

199. **Humidity.**—Review the section carefully. Observe that air at any given temperature can hold a certain amount of water vapor. If you raise the temperature it will hold a larger amount of water. Whenever a body of air has all it can hold at the given temperature, we say it is "saturated," and we call the amount of water present the *absolute humidity*. But suppose the air has only one-third the water vapor it would hold at its present temperature, we then speak of its relative humidity as 33⅓ per cent—that is, the ratio between its contents and its

capacity.* This is the important fact about the moisture of the air in our study of weather and climate, and the student should master the idea.

How relative humidity is determined. Two thermometers are used; one the ordinary instrument, and the other, one whose bulb is wrapped in muslin, which is kept wet from a receptacle containing water by an arrangement similar to a lamp-wick. These instruments are exposed in the same place, and are known as wet-bulb and dry-bulb thermometers. Water evaporating from a surface always takes away heat, as when the body dries without rubbing after a bath. Thus the evaporation going on from the surface of the wet muslin cools the bulb, and the instrument marks a lower temperature than the other. If now the air has much less moisture than it will hold, it will act like a dry sponge, and the evaporation and consequent cooling of the wet bulb will be large. If the air, on the other hand, is almost saturated, it will act like a wet sponge, and the evaporation and cooling in case of the wet bulb will be slight. The amount of cooling, or the difference between the readings of the wet-bulb and dry-bulb thermometers, becomes therefore a measure of the moisture in the air. Physicists have thus been able to construct tables which will tell the relative humidity for any given temperature of the air, when the difference of the readings described above is known. Such tables are in use everywhere by students of meteorology. An abbreviated table sufficient for our study is given on the next page. For full tables refer to Ward's Practical Exercises in Elementary Meteorology, pages 146–165.

---

* It is well to observe that strictly it is the *space* and not the air which holds or contains the water vapor.

TABLE FOR DEW-POINT AND RELATIVE HUMIDITY.

DEW-POINT TABLE.

Difference between Wet- and Dry-Bulb Readings (degrees).	TEMPERATURE OF AIR.											
	-10°	0°	10°	20°	30°	40°	50°	60°	70°	80°	90°	100°
2	−76	−18	−1	+12	+24	+35	+46	+57	+67	+77	+87	+98
4	....	....	−22	+1	17	30	42	53	64	74	84	95
6	....	....	....	−18	7	24	37	49	61	72	82	93
8	....	....	....	....	−8	16	31	45	57	68	79	90
10	....	....	....	....	....	4	25	40	53	65	77	87
12	....	....	....	....	....	−16	17	35	49	62	74	85
14	....	....	....	....	....	....	5	28	45	58	70	82
16	....	....	....	....	....	....	−20	20	39	54	67	79
18	....	....	....	....	....	....	....	8	33	50	63	76
20	....	....	....	....	....	....	....	−13	25	45	60	73
22	....	....	....	....	....	....	....	....	15	39	56	69
24	....	....	....	....	....	....	....	....	0	32	51	66

RELATIVE HUMIDITY.

Percentages.

Difference	-10°	0°	10°	20°	30°	40°	50°	60°	70°	80°	90°	100°
2	10	42	60	72	79	84	87	89	90	92	92	93
4	....	....	21	44	58	68	74	78	81	83	85	86
6	....	....	....	16	38	52	61	68	72	75	78	80
8	....	....	....	....	18	37	49	58	64	68	71	74
10	....	....	....	....	....	22	37	48	55	61	65	68
12	....	....	....	....	....	8	26	39	48	54	59	62
14	....	....	....	....	....	....	16	30	40	47	53	57
16	....	....	....	....	....	....	5	21	33	41	47	51
18	....	....	....	....	....	....	....	13	26	35	41	47
20	....	....	....	....	....	....	....	5	19	29	36	42
22	....	....	....	....	....	....	....	....	12	23	32	37
24	....	....	....	....	....	....	....	....	6	18	26	33

Observe that the figures at the top give the temperature as shown by the dry-bulb thermometer—that is, the actual temperature. The difference in the two readings appears in the left-hand column. Suppose you find the difference 6° and the temperature is 70°. The relative humidity is therefore 72—that is, the air has 72 per cent, or not quite three-fourths of the moisture which it will hold. In this case the relative humidity is high.

But suppose now the difference in readings is 20°. This means that much evaporation is taking place and much cooling. Let the dry-bulb reading be 70°, as before. Now our relative humidity is only 19—that is, the air holds less than one-fifth of the moisture for which it has capacity.

The two thermometers are often mounted together and constitute the instrument known as the psychrometer, a *measurer* of *vapor*. As directed above, use the psychrometer and the table in finding the relative humidity for a series of days, and record it in your note-book.

Also make a series of observations in one day, choosing a day when variations of temperature from morning until midday and from midday to evening are considerable. What variations in relative humidity have occurred during the day? Has there been necessarily any change in the actual amount of moisture (absolute humidity)? If the amount of moisture was constant or nearly so, why did the air feel much more damp after sundown than at two in the afternoon? Dr. Hann, a German meteorologist, cites a relative humidity of 9 per cent about the middle of August in the heart of the Libyan desert. But he adds that the amount of vapor in the air was equal to that of the damp winter air of Western Europe, which shows a relative humidity of 80–90 per cent. Explain this. In what sense is the same air with the same amount of water vapor dry at one time and wet at another? In connection with your answer, we may quote again from Dr. Hann, who says that the Libyan desert conditions represent "typical desert dryness, in which finger nails split and the skin peels off." * Why in the old days were apples

* Handbook of Climatology, translated by Ward, p. 51.

quartered, strung, and hung above the stove? Is the principle different from that of modern evaporation processes? Hops are dried in an upper chamber on a canvassed open floor above a furnace. Why is a ventilator always placed at the top of the kiln? Why do the clothes on the line dry better on some freezing days than on some warm days?

In Vienna the average relative humidity in January for twenty years was as follows:

6 A. M.	2 P. M.	10 P. M.
87	77	86

The average for corresponding hours in July was—

75	48	70

The actual amount of water in the air (absolute humidity) was three times as great in July as in January. Explain the low relative humidity at 2 P. M. in July, and the similar relative humidities of 6 A. M. in the different months.

200. **Dew and Frost.**—Read the section and fix the principle on which the gathering of the dew depends. It will thus be plain that the dew-point may be reached in two ways: by increasing the humidity, the temperature remaining the same; or by lowering the temperature, the humidity remaining the same.

Select any cold winter day. Fill a kettle with water and boil it until it steams freely for some time. A mist gathers on the window-pane (if very cold, a frost). Which method of dew formation does this illustrate? Why is the dew likely to gather on a still, clear night after a rain? Does the above experiment tell you the temperature at which the dew-point is reached?

Pour a small quantity of water at an ordinary temperature into a tin basin or cup. Add ice-water or small pieces of ice and stir with a thermometer. What change after a time takes place on the outside of the cup? When this happens read the thermometer. It give the temperature of the water. But the water has given its own temperature to a film of air close to the cup. At this temperature the air has given up some of its moisture—that is, dew has formed. You have the temperature of dew formation for that particular mass of air, or the dew-point. Thus we have illustrated the second method of dew formation given above.

We need not actually create dew in order to learn the dew-point at any particular time and place. Dew is formed when the relative humidity becomes 100 per cent, or when the air is saturated. Physicists can determine the temperature at which this takes place if they know their relative humidity. Consult the dew-point table. Suppose that the temperature is 70° and the difference between wet and dry bulb readings is 6°. The table gives the dew-point as 61°. Determine the dew-point for the same time and place by each of the methods outlined and see if the results tally.

If the temperature of the saturated air falls below 32° F. a "white frost" appears instead of dew. Why is this more abundant in spring and autumn than in winter? Does the *white frost* or the *low temperature* harm vegetation? Why is frost not to be expected in a cloudy night in spring or autumn? Why would a smudge be effective beyond the reach of the heat-rays from the fire? In what way, precisely, does the covering of flowers or vegetables protect them from late or early frosts?

Answer if you can such questions as these: Why are frosts less likely to occur just after a rain? Why are frosts common on low ground, while slopes or hilltops may escape? Why is there less danger of frosts when the night is cloudy? If you have not thought of these questions, it would be well to notice and record the exact conditions of the soil, of the wind, of the sky, and of the topography in case of the spring and autumn frosts. If you can not think out satisfactory answers to the questions given above, seek them from reading and in the discussions of the class-room.

The Weather Bureau makes much of frost warnings, which have had great value in leading to the protection of fruit and other crops, as in California, where frosts are said to be the severest enemy of crops, excepting the insect pests. Thus March and April frosts injure almonds, apricots, peaches, grapes, and prunes, and the citrus fruits are damaged by frosts in December, January, and February. Various methods of protection are employed, such as screens of cloth or laths. But this is impracticable to any large extent, and other means are used, as irrigation, before a frost is expected, smudge fires with damp fuel, plants for evaporating water in the fields, or direct warming of the air. The student should seek to explain why each of these methods may be effective.

201. **Clouds and Fog.**—The student should study first the "state of the sky." Some other expression might be used, but this is commonly employed, of the conditions of the sky as regards clouds and sunlight. Look carefully around the horizon and up to the zenith and estimate the proportion of the expanse covered by clouds. Your estimate will be rough, but you can, by practice, come close to the truth. If there are no clouds, you will of course record the sky as *clear*. If there are clouds, covering less than $\frac{3}{10}$ of the expanse, the sunlight will still prevail, and you will call the sky clear. If more than $\frac{3}{10}$, but less than $\frac{7}{10}$ are clouded, you will place *fair* in the record. It is not strictly true, but this technical sense is given the word, so that we may have language upon which all students may agree, and thus be able to understand each other. If more than $\frac{7}{10}$ of the sky is covered, you will record it as *cloudy*. Continue your record for a month

and then summarize, recording respectively the number of clear, fair, and cloudy days.

The classification of clouds. As the atmosphere is a great ocean of gases, with more or less water vapor in an invisible or visible state, we should expect to find it in a condition of change, and clouds in many forms will appear and disappear, at high levels and low, over ever shifting areas. These changes will depend on the temperature of the surface, on winds and up-drafts, on relations of land and sea, and on the forms of the land. The resulting cloud forms are many, and they constantly change and grade into each other. Of the many types recognized, the student should learn at least three or four of the more distinct and common. Begin by referring to Fig. 157. What is the name of these clouds? How are they described on page 230? Are they high or low clouds? If no such clouds are now visible, watch for them and describe their appearance as fully as you can. If you have access to Illustrative Cloud Forms, published by the United States Hydrographic Office, study plate i and the accompanying definition. Consult the Century, Standard, or other dictionary for the meaning of the word cirrus. In the same way look for cumulus and stratus.

What clouds are shown in Fig. 158? Do you recall the mountainous appearance of "thunder-heads" seen in thunder-storms? Have you seen some clouds to which the term cumulus applies more exactly than it applies to Fig. 158? What is the difference between the clouds of Fig. 158 and Fig. 159? Compare Illustrative Cloud Forms, plate viii. What do you think of the term "wool-pack," as applied to cumulus clouds? See also plate xiii.

Have you observed clouds in narrow, horizontal bars, and lying low in the sky? These are examples of stratus. When surrounded by fog, you are simply in the midst of a stratus cloud which rests on the earth's surface? Observe now that cirrus is a high cloud, that cumulus may be high or low, while stratus including fogs is low. You will often be unable to place clouds under one of the above heads: Cirrus clouds may have somewhat the appearance of stratus—mackerel sky—or cumulus (Illustrative Cloud forms, plates ii and iii), and cumulus may resemble stratus (Illustrative Cloud Forms, plate vi). If you can not classify, describe the forms as they appear to you. Remembering that cirrus clouds average several miles in altitude, can you explain the fact that they consist of small ice crystals? Which of the cloud-forms above studied could be formed by up-drafts of cooling and condensing air? What sometimes results from the formation of such clouds? See section 227. Suppose the air close to the earth cools by radiation during the night below the dew-point; what kind of a cloud will be present in the morning?

Observe changes in the form or position of the clouds. Do any clouds appear to be moving under the influence of winds? If so, in what direction? Do you notice movement in high clouds, when no wind is blowing where you are? Do the clouds seem to be changing form without moving from one position to another? Can a cloud change its position without motion of its constituent particles of water? How could this happen? Could a cloud remain stationary while its particles were changing? The cloud banner stretching in the direction of the wind from the Matterhorn and other lofty peaks is an example.

Record the history of any particular day as regards cloudiness. If the day begins clear and becomes partially or wholly cloudy, observe and record the succession of forms, the rate at which they developed, the quarter of the sky where they began. Observe any connection that the cloud history of the day appears to have with its winds, its changes of temperature, its rain- or snow-fall.

202. **Rain and Snow.**—These belong under the general head of *Precipitation,* which also includes hail and even dew. The total amount of the latter two is so slight as to be neglected in ordinary measurements. Record the character and the changes of any single rain-storm. Does it begin suddenly or gradually? What are the attendant conditions of cloud and wind? How long does it last? How much water falls? To answer the last question a rain-gage is necessary. Any vessel with vertical sides would serve for a few observations. It should be placed where the fall of rain is unobstructed by buildings or other objects. Measure the depth with a ruler. The fall is often so slight as to make accurate results impossible, hence the rain-gage as used by weather observers is so made that the water as it falls is concentrated in a cylinder of smaller cross-section than that into which it first falls, and is thus ten times deeper than the layer of water would be if it remained spread out on a level surface. The measuring rule is so graduated as to compensate for this and enable the observer to read off at once the inches and hundredths of an inch. Snow falling into the receptacle is first melted and then gaged in the ordinary way. When snows fall without wind upon a smooth field the thickness may be measured and recorded as 1 inch (water) for each 10 inches of snow. Melt snows from different falls and see how nearly this rule approaches to accuracy. If you have carried on this observation for a month, give a general account of the precipitation for the period. How many rain- or snow-storms? In how many was the amount less than ½ inch? Between 1 inch and ½ inch? More than 1 inch? Russell states that most rainfalls do not amount to more than 0.5 of an inch; that falls of 1 inch are common, while greater falls are less frequent. The extreme record in the United States for twelve years gives 22.3 inches in one period of twenty-four hours, June

15–16, 1886, at Alexandria, La. Russell gives a few still more surprising records for India, France, and Italy.*

203. **Rainfall.**—Refer to Fig. 160. · What symbol is used for areas having over 60 inches of rainfall? For areas having 30 to 40 inches? 5 to 10 inches? In general, how many areas of large rainfall in the United States? Where are they, and what States are largely included? How many areas have over 60 inches? Where is the most westerly of these? Where is the most southerly? Describe the extent and form of the large area which lies but partly in Florida. What other areas in the Eastern United States? Name the three States which share in one of these. Records for two years at Horse Cove, North Carolina, in this area, give 93.33 inches of annual rainfall. Describe the western area, giving the States in which it lies? Observe a small area of this shade in California. Is the land at that point high or low? How many of the regions of more than 60 inches of rainfall are on the sea-border? Has the sea any important effect on the others? The highest rainfall in the United States is in Washington. At Neah Bay in that State records of eight years show 108 inches annual average, with maximum of 136 inches in 1879. How does this compare with the delta region of the Ganges? (See page 235.) What is the rainfall of the Northeastern States? Does the great belt of 30 to 40 inches touch the Atlantic? The Gulf? What is the general direction of its western boundary? The extent (in length), the width, and direction of the belt of 20 to 30 inches? What States are crossed by it? Is any State nearly covered by it? What can you say of the agriculture of the 30- to 40-inch belt? Is irrigation necessary in the 20- to 30-inch belt?

Where is the greatest east and west extension of the 10- to 20-inch belt? The greatest north and south exten-

---

* Meteorology, Thomas Russell, p. 93.

sion? Where do you find islands of greater rainfall in this belt? What is the cause of them? Is irrigation always needed in the 10- to 20-inch belt? Is it common? Describe the areas of less than 10 inches. Have any less than 5 inches? Below what standard of rainfall do we apply the term "arid"? What is the general explanation of the north and south wet and dry belts in the Western United States? What is the rainfall of eastern Washington and Oregon? Observe that these States have heavy rainfall and heavy frosts in the west, and much necessity for irrigation in the east. Oregon averages 44 inches and Washington 39.8 inches. This is not very different from Pennsylvania. Let the student remember that these are averages for whole States, and also consider the extremes in both cases, and the important effects on agriculture.

Rainfall of California.—See Fig. 14 of this manual. Review what you have learned of the topography of California. See relief map, Fig. 129. What is the highest precipitation shown in the rainfall map? What is the symbol used? How far does it follow the shore-line? How far does it appear along the Sierras? Describe the 30- to 40-inch belt in relation to the Great Central Valley and the mountain ranges of the State. Would the 20- to 30-inch belt and the 10- to 20-inch belt appear in like manner continuous if a rainfall map of Nevada were supplied? What is the rainfall at San Francisco? How far does this range of precipitation extend along the shore-line as seen in this map? How do the northern and southern parts of the central valley compare in precipitation? What term should be applied to the southern area? Why do we find a wet belt followed by a dry belt twice as we cross the State to the east? Why do we find the State progressively more arid as we go southward? The student should note that in southern California, in the Colorado desert, are places where the annual rainfall is but 2 inches, or even less. On the other hand, the belt having more than 40 inches is very imperfectly shown on the map,

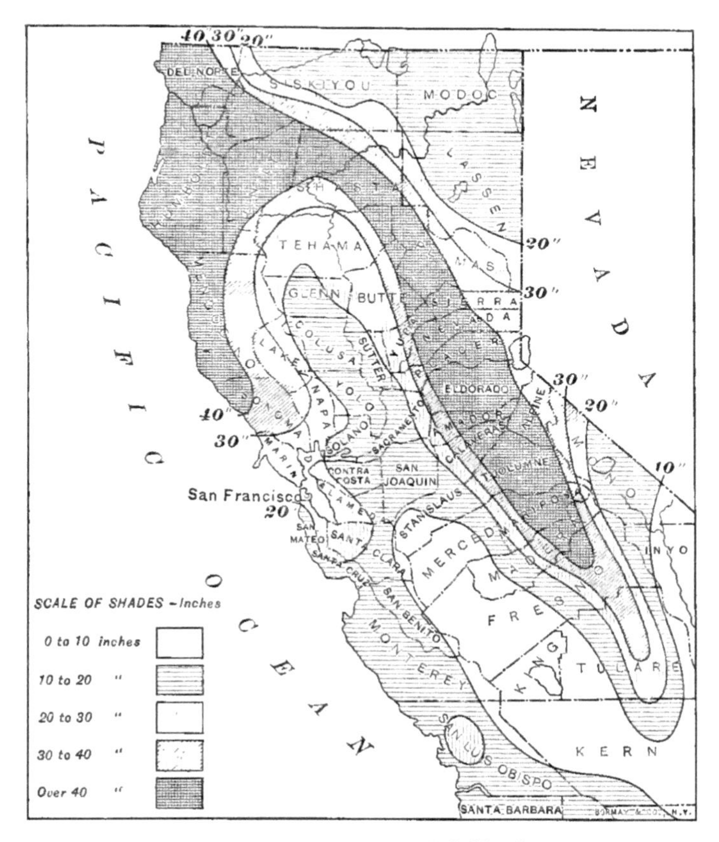

FIG. 14.—Rainfall map of California.

for within it are points averaging 75 inches, and individual year records showing 102, 111, 119, and 120 inches. Write a single paragraph, stating the influence of the ocean and of the topography on the rainfall of the State.

207. **Measurement of Temperature.**—The principle of the thermometer should be mastered from section 207 and the work of the class-room. It would be useful to begin work in your note-book with a careful drawing of the instrument which you are to use. Your thermometer should be hung in the shade, on the north side of the house, or under a cover at a short distance from any building. It should be kept dry, and should not be handled while reading the temperature. State the reason for each of these precautions.

Most persons read the thermometer because of interest in very low or very high temperatures, or to see whether a frost is probable. But the student of the atmosphere has a broader purpose—namely, to learn the history of the temperature, through a day, through a succession of days, through several seasons, and for a period of years. He will thus learn the average or *mean* temperature of the locality and the range of its upward and downward extremes. Could you gain this knowledge by reading your thermometer but once in each day? Why? Select three hours: 7 A. M. (if this is too early, adopt 8 A. M., but the former is better); 2 P. M. and 9 or 10 P. M. Observe that you have two hours which are neither the hottest nor the coldest of the twenty-four-hour period, and one, 2 P. M., which is the warmest time (usually) of the day. Why at 2 P. M. rather than at noon? Experience shows that by adding the three records and dividing by three we get with fair accuracy the mean temperature. A closer approach to the true mean is obtained by using the following rule: Add the 7 A. M., the 2 P. M., and twice the 9 P. M. temperature and divide by 4. With coördinate paper make a graphic record, like that in Fig. 166, continuing for at least one full month. Add the averages for all the days and divide by the number of days in the month. What is the resulting figure? Which was the warmest day? The coldest? Describe the "curve" of temperature. If your record continues two or three months compare the averages and the extremes. If you were to continue the record for twelve months, how would you get the mean temperature for the year? Could you get the mean temperature for a month if your observation was confined to school hours? Could you get the mean temperature for a year if you made a record only during the months of school?

Many thermometers, especially the cheaper ones, do not give true readings. They may be nearly correct for ordinary temperatures, but much in error for extremes. The student may test the freezing-point of an instrument by mixing water and crushed ice and immersing it in the water thus brought to the freezing-point. Similar tests may be made of the boiling-point, observing the precaution that if the tube is too short the mercury will burst it by expanding with the heat. It is a useful exercise to test different parts of a room for temperature, as near the floor and ceiling, or against an outside as compared with an inside wall.

Review the meaning of *conduction* and *radiation* as given in section 209. We may also note another way of warming or cooling: when warming or cooling is accomplished by actual circulation or movement of air from one position to another, the process is called *convection*, which means carrying. We use various means to increase the warmth of our houses in winter, among them being storm-doors, extra windows, dampers in fireplace chimneys, paper linings in the walls, packed floors, and wind breaks. What process of losing heat is checked in each of these cases? The answer can not be exact in all cases, as in that of a wall through which loss occurs both by conduction and radiation.

207*a*. **To Convert Centigrade Degrees to the Fahren-**

heit Scale.—Multiply by 1.8 and add 32. To change Fahrenheit to Centigrade, therefore, subtract 32 and divide by 1.8.

At Yarmouth, on the east coast of England, the mean temperature for January is 2.9° C.; and for July, 15.9°. What is the temperature by the Fahrenheit scale? Compare the results with the isotherms crossing Great Britain, as seen on the maps, Figs. 170 and 171. What is the difference between the January and July temperatures, Fahrenheit? At Halifax, Nova Scotia, the January mean is −5.5° C. and the July mean is 17.7° C. How does the difference or "range" compare with that of Yarmouth, England?

The mean January and July temperatures of Chicago from 1881 to 1890 are given (Fahrenheit) in the following table. The January mean for Vienna is —1.5° C. and the July mean is 19.8° C. Compare the annual range of the two cities, using the mean for ten years in Chicago.

	January.	July.
1881	19.5 deg.	72.9 deg.
1882	28.3 deg.	68.6 deg.
1883	16.3 deg.	71.0 deg.
1884	19.2 deg.	69.2 deg.
1885	18.3 deg.	72.8 deg.
1886	21.4 deg.	71.4 deg.
1887	17.3 deg.	76.0 deg.
1888	15.1 deg.	72.6 deg.
1889	29.0 deg.	70.5 deg.
1890	30.8 deg.	72.1 deg.

216. **Temperature Maps—Isotherms.**—Review section 216 and give further study to Fig. 169. What is the temperature of Denver? Lincoln? Philadelphia? and Indianapolis? What places have temperatures between 10° and 20°? Between −10° and −20°? Why do the isotherms

bend southward in Montana, Wyoming and Colorado? Why do they bend northward in the Upper Mississippi Valley? Why should New England be warmer than northern and western New York? What share of the United States (by estimate) had temperature above the freezing-point? If some territory along the Pacific coast (the lines not being here shown) was above freezing, could the area have been large? Why not? If an isothermal for 32° had been drawn, near what cities would it probably appear? What general effect is produced by the ocean waters on the east, south, and north? Name all the important influences which control the distribution of temperature, as suggested to you by the study of this map.

216*a*. **Isotherms for January and July.**—Figs. 170 and 171. Take first the January map. Describe from west to east the course of the heat equator. Why is it so generally south of the equatorial line? Why is it north of this line in the eastern Pacific Ocean? Why on the coast of Guinea in West Africa? Does the heat equator approach the Equator at any other point? Observe the areas having a temperature of 80° or more. Do they form a continuous belt around the globe? Why do the heat equator and the isothermals of 70° and 60° bend so far south upon crossing the western shore-line of South America? Is the passage from ocean to the land enough to explain it? Compare the course of the same lines as they pass from the Atlantic to the east coast of the same continent. Where, on the waters, are the isothermals most nearly straight? Why? Where on the lands? Why? Study the isothermal of 30° in the northern hemisphere. Describe its course from west to east. What States in the Eastern United States have the same January temperatures as southern Alaska and the Aleutian Islands? In about what latitude is the most southerly position of this isothermal? What is its most northern point?

How does the temperature in the Black Sea region compare with that of western Scandinavia? What causes the chief curvature of the lines in the United States?

Taking the map for July, trace the heat equator from west to east. Does it anywhere run south of the Equator? Why is the heat equator not pushed southward in July as effectively as it is pushed northward in January? Where, in the equatorial region, is there a great break in the belt of 80° or more? Where is the 80° belt very narrow, and why? Why do the isothermals of July bend so far northward in North America? Why do they run so directly across the North Atlantic as compared with their course in January?

Let the student also examine the isothermals for the year, Fig. 172. Trace the heat equator from west to east. Why does it average about 10° north of the geographic equator? Compare the isotherm of 40° with those of January and July, all in the northern hemisphere.

216*b*. **The Making of a Temperature Map.**—The following table gives data for a complete weather map of the United States:

DATA FOR WEATHER MAP OF THE UNITED STATES, FOR 8 A.M. (EASTERN TIME), MARCH 1, 1902.

Place.	Pressure in inches.	Temperature.	Wind.		Precipitation in previous twenty-four hours (inches).
			Miles per hour.	Direction.	
Atlantic Coast States and St. Lawrence Valley					
Sydney, C. B. I.	30.20	32	8	S.	Trace
Father Point, Quebec	29.50	36	16	S. E.	.10
Quebec, Quebec	29.36	40	24	N. E.	.56
Eastport, Me.	29.66	44	34	S.	.52
Northfield, Vt.	29.48	44	12	N.	.64
Portland, Me.	29.52	42	20	S.	2.12
Boston, Mass.	29.62	48	16	S. W.	1.72
Nantucket, Mass.	29.74	46	20	S.	.60

PLACE.	Pressure in inches.	Temperature.	WIND.		Precipitation in previous twenty-four hours (inches).
			Miles per hour.	Direction.	
Block Island, R. I.	29.70	44	22	S. W.	1.34
Binghamton, N. Y.	29.52	44	Lt.	S.	.96
Albany, N. Y.	29.58	46	8	S.	1.36
New York, N. Y.	29.64	48	10	S. W.	.44
Scranton, Pa.	29.58	44	Lt.	N. E.	1.60
Philadelphia, Pa.	29.64	52	8	S.	1.58
Atlantic City, N. J.	29.72	42	14	S. W.	.58
Baltimore, Md.	29.62	50	Lt.	S.	.82
Washington, D. C.	29.60	50	6	S.	1.08
Lynchburg, Va.	29.62	50	Lt.	S. W.	.04
Norfolk, Va.	29.72	54	12	S.	.62
Charlotte, N. C.	29.68	46	8	S.	.04
Raleigh, N. C.	29.72	54	10	S. W.	.32
Hatteras, N. C.	29.80	52	20	S. W.	.38
Wilmington, N. C.	29.78	58	12	S.	.90
Charleston, S. C.	29.76	56	8	S.	.01
Augusta, Ga.	29.68	58	Lt.	S.	0
Savannah, Ga.	29.72	60	8	S. W.	.02
Jacksonville, Fla.	29.74	60	12	S.	.0
Jupiter, Fla.	29.72	68	Lt.	N.	.14
Key West, Fla.	29.80	74	Lt.	S.	.14
Gulf States					
Atlanta, Ga.	29.62	48	14	S.	.04
Macon, Ga.	29.60	58	8	S.	0
Tampa, Fla.	29.76	60	8	E.	.08
Mobile, Ala.	29.60	60	Lt.	S. E.	0
Montgomery, Ala.	29.60	56	18	S. W.	.04
Meridian, Miss.	29.52	58	16	S. W.	Trace
Vicksburg, Miss.	29.50	56	6	S. W.	0
New Orleans, La.	29.56	62	8	S.	0
Shreveport, La.	29.52	52	6	W.	0
Fort Smith, Ark.	29.50	38	8	S. W.	.01
Little Rock, Ark.	29.48	38	10	S. W.	Trace
Palestine, Texas.	29.58	54	6	N.	0
Galveston, Texas.	29.52	62	14	S. W.	0
Taylor, Texas.	29.58	48	8	W.	0
San Antonio, Texas.	29.64	56	6	S. W.	0
Corpus Christi, Texas.	29.64	58	8	N. W.	0
Ohio Valley and Tennessee					
Memphis, Tenn.	29.48	44	14	S.	.02
Nashville, Tenn.	29.50	44	10	S. W.	.06
Chattanooga, Tenn.	29.56	46	6	S.	.01
Knoxville, Tenn.	29.54	46	10	S. W.	Trace

PLACE.	Pressure in inches.	Temperature.	WIND.		Precipitation in previous twenty-four hours (inches).
			Miles per hour.	Direction.	
Louisville, Ky.	29.40	42	14	S.	.14
Evansville, Ind.	29.34	40	14	S.	Trace
Indianapolis, Ind.	29.28	36	22	S. W.	.14
Cincinnati, Ohio	29.38	44	12	S. W.	.14
Columbus, Ohio	29.36	42	12	S. W.	.10
Parkersburg, W. Va.	29.44	50	10	W.	.02
Pittsburg, Pa.	29.40	48	10	S.	.18
Lake Region					
Bissett, Ontario	29.32	34	0	....	.10
Port Arthur, Ontario	29.04	10	22	W.	.28
Parry Sound, Ontario	29.18	40	12	S. E.	.18
Saugeen, Ontario	29.20	38	16	S.	.06
Oswego, N. Y.	29.44	46	14	S.	.70
Rochester, N. Y.	29.42	44	8	S.	.22
Buffalo, N. Y.	29.34	.46	28	S.	.16
Erie, Pa.	29.30	48	20	S. W.	.10
Cleveland, Ohio	29.30	44	24	S.	.02
Toledo, Ohio	29.24	40	18	S.	.04
Detroit, Mich.	29.22	38	20	S. W.	.06
Alpena, Mich.	29.08	36	14	S. E.	.14
Sault Ste. Marie, Mich.	29.00	38	10	S. E.	.30
Houghton, Mich.	28.88	26	Lt.	E.	.62
Marquette, Mich.	28.88	32	6	S. W.	.48
Escanaba, Mich.	28.92	32	12	S.	1.12
Green Bay, Wis.	28.96	32	12	S. W.	.82
Grand Haven, Mich.	29.04	34	26	S.	.44
Milwaukee, Wis.	29.02	34	10	S. W.	.12
Chicago, Ill.	29.10	34	24	S. W.	.08
Duluth, Minn.	29.06	16	24	N. W.	.14
Upper Mississippi Valley					
St. Paul, Minn.	29.06	18	12	N.	.26
La Crosse, Wis.	29.02	24	10	W.	.12
Dubuque, Iowa	29.10	30	8	W.	.46
Davenport, Iowa	29.08	30	16	W.	.18
Des Moines, Iowa	29.22	24	18	W.	.10
Keokuk, Iowa	29.22	28	18	S. W.	.16
Springfield, Ill.	29.24	32	20	W.	.08
St. Louis, Mo.	29.30	36	24	S. W.	Trace
Cairo, Ill.	29.40	40	12	S. W.	.01
Missouri Valley					
Springfield, Mo.	29.38	32	16	N. W.	.04
Kansas City, Mo.	29.42	26	12	N. W.	.08

Place.	Pressure in inches.	Temperature.	Wind.		Precipitation in previous twenty-four hours (inches).
			Miles per hour.	Direction.	
Wichita, Kan.	29.54	34	14	N. W.	0
Concordia, Kan.	29.50	32	24	N. W.	0
Omaha, Neb.	29.34	22	20	N. W.	.08
Valentine, Neb.	29.58	20	24	N. W.	Trace
Sioux City, Iowa.	29.30	20	30	N. W.	.08
Huron, S. Dak.	29.46	14	36	N. W.	.12
Pierre, S. Dak.	29.64	22	30	N.	.20
Moorhead, Minn.	29.42	8	28	N. W.	.16
Bismarck, N. Dak.	29.76	10	34	N. W.	0
Williston, N. Dak.	29.98	10	12	N.	.30
Rocky Mountain Slope					
Battleford, Sask.	30.02	4	Lt.	S. E.	0
Havre, Mont.	29.96	22	Lt.	S. W.	Trace
Helena, Mont.	29.92	32	Lt.	S. W	0
Miles City, Mont.	29.92	24	6	N. W.	0
Kalispell, Mont.	29.90	30	Lt.	W.	.02
Lewiston, Idaho.	30.00	36	Lt.	....	Trace
Pocatello, Idaho.	30.04	28	6	S.	.84
Boise, Idaho.	30.08	32	Lt.	S. E.	.02
Rapid City, S. Dak.	29.80	24	22	N. E.	.02
Lander, Wyo.	29.98	12	Lt.	S.	.06
Salt Lake City, Utah.	30.06	30	Lt.	E.	.22
Modena, Utah.	30.00	22	Lt.	N. W.	Trace
Grand Junction, Col.	29.98	24	Lt.	E.	.10
Cheyenne, Wyo.	29.78	22	26	N. W.	Trace
North Platte, Neb.	29.50	28	28	N. W.	0
Denver, Col.	29.74	26	Lt.	N.	0
Amarillo, Texas.	29.70	30	12	N. W.	0
Pueblo, Col.	29.70	28	8	W.	0
Dodge, Kan.	29.60	32	14	N. W.	0
Oklahoma, Oklahoma.	29.56	36	8	N. W.	0
Fort Worth, Texas.	29.64	42	8	N. W.	0
Abilene, Texas.	29.68	44	12	N. W.	0
El Paso, Texas.	29.80	38	8	W.	Trace
Santa Fé, N. Mex.	29.78	24	8	N.	.01
Flagstaff, Ariz.	29.94	32	Lt.	N.	0
Yuma, Ariz.	29.94	42	Lt.	E.	0
Phœnix, Ariz.	29.94	40	0	....	0
Pacific Coast					
Barkerville, Brit. Col.	29.74	20	0	....	0
Victoria, Brit. Col.	29.92	40	0	....	.08
Kamloops, Brit. Col.	29.76	36	0	....	0
Spokane, Wash.	30.00	34	8	S.	.16

Place.	Pressure in inches.	Temperature.	Wind.		Precipitation in previous twenty-four hours (inches).
			Miles per hour.	Direction.	
Tacoma, Wash.	29.98	42	6	S. W.	.20
Portland, Ore.	30.00	42	8	S.	.42
Roseburg, Ore.	30.02	34	Lt.	E.	.12
Baker City, Ore.	30.04	24	Lt.	S. E.	.04
Carson City, Nev.	30.00	22	Lt.	N.	0
Winnemucca, Nev.	30.04	22	Lt.	N. E.	0
Eureka, Cal.	29.96	42	Lt.	S. W.	.32
Red Bluff, Cal.	29.96	42	Lt.	N.	0
San Francisco, Cal.	29.94	48	8	S. E.	0
Fresno, Cal.	29.98	58	Lt.	N. E.	0
Los Angeles, Cal.	30.02	48	8	N. E.	0
San Diego, Cal.	29.98	46	Lt.	N. E.	0

Use at this time only the figures in the third column, those giving temperature. On the blank weather map record the temperature of each place as given. Make the figures small and neat, and place them close to the small circle marking the position of the town (location of weather station). What place has the highest temperature of any recorded here? What have a temperature of 60°? What places have a temperature of 10°? Are any below this point? In what part of the United States are most of the stations showing between 20° and 30°? Compare the temperatures of eastern Washington, Kansas, and the region about Lake Michigan. How much of the Atlantic shore-line has temperatures between 40° and 50°? Between 50° and 60°? Between 60° and 70°? What is the range of temperature on the Atlantic shore-line? On the Pacific shore-line?

We may call the map as we now have it a temperature map, but it requires considerable study in detail to see how the warmer and colder places are distributed. We can make a more useful and expressive map by drawing lines of equal temperature or isotherms. Let the student use a pencil so that mistakes can be corrected.

What place has a temperature nearest approaching that of Key West? If the temperature increases at a uniform rate from Jupiter to Key West, you would find 70° about one-third of the distance from the former to the latter. We may therefore draw a line across southern Florida through the probable point of 70°. All points north of it will have, so far as our record shows, temperatures below 70°; all south of it, about 70°. Notice the temperature of Corpus Christi and of Galveston. Places having 60° would naturally be expected a little south of the former and a little north of the latter. Note the temperatures of Taylor and Palestine, and determine how far north of Galveston the isothermal of 60° should pass. What places have a record of exactly 60°? The line must of course pass through these. Where will the line run between New Orleans and Vicksburg? How far to the northeast must you go to carry the line from Mobile to Tampa? Draw the line on easy curves, and not straight from point to point. What share of the Gulf coast of the United States is included within the belt of 60°? Might there be warmer places of which we have no record? Draw the isothermal of 50°, leaving all places of higher temperature to the southeast and all places of lower temperature to the northwest. Where does the line break on the east? At the southwest? Could we carry it out in both directions if we had the records of temperatures for the Atlantic and for Mexico? Does the line of 50° come in again on the Pacific slope? Observe the temperatures of the California stations. Imagine yourself carrying a thermometer from Fresno City eastward, southward, or toward San Francisco. Could you go in any of these directions without meeting a temperature of 30°?

In like manner draw the line for 40°. How is Lake Ontario situated with reference to it? Describe its curves from Phœnix, Ariz., to Quebec. Where does it come to the Pacific shore-line? In like manner draw the lines for 30°, 20°, and 10°. Compare the temperatures in going from Springfield, Mo., to Bismarck, Dak., with temperatures and distances in passing from Marquette, Mich., to Port Arthur, Canada. What point in the Rocky Mountain region has a temperature below 20°? If you go from this place to Rapid City, Cheyenne, Salt Lake City, or Pocatello will you probably find temperatures of 20°? Draw the proper isothermal, including this island of low temperature. Observe in the same manner an island of higher temperature at Great Salt Lake. Could you draw the line for 32°? What share of the United States (by estimate) was below freezing on the date of these observations? How do the isothermals of 40° and 50° run with reference to the Atlantic and Gulf coasts? In answering this question leave Florida out of account, and think of a single shore-line from Texas to Nova Scotia. Compare the temperatures of New York and Tennessee; also those of Michigan and Arizona; that of Boston and of Lander, Wyo. Note also San Diego and Victoria. What influences occur to you as modifying the temperatures that might be expected from the several latitudes? With water-color or crayon, color the belt of temperature to which your home locality belongs.

216*c*. **Temperature Maps for Successive Days.**—Turn to Fig. 174. This, like Fig. 169, belongs to January 7, 1886. Disregard all weather signs except the isotherms, and compare with Fig. 175, noting dates. Does the map for January 8th show higher temperatures at any point? Has the Montana area of low temperature shifted its position, and how? How has the temperature of Chicago changed? Has the temperature of New York City changed more or less than this? What was the temperature of central Texas on the morning of January 7th?

On January 8th? On January 9th? What change in the temperature of North Dakota from January 8th to January 9th? Compare the temperatures of the other stations or regions for the three days.

Study the isotherms of the current weather maps for one week, and record the principal changes in the distribution of temperature from day to day. Color the belt of some selected range of temperatures for the week, and place the maps together in their proper order for a comparative view.

218. **Pressure of the Atmosphere—The Barometer.**—Review the section so that the principle of the barometer is clearly in your mind. It will aid to fix it in memory if you perform the following simple experiment: Fill a test-tube with water, stop the open end with your finger and invert it, placing the open end in a basin of water before removing your finger. The air, pressing on the surface of the water in the basin, prevents the water from falling in the tube. If your tube of water were 34 feet high the weight of the air would just balance it. That is, the column of water of that height is equal in weight to a column of air extending from the earth's surface to the upper limit of the atmosphere. Mercury is so much heavier that a column about 30 inches high will balance the air pressure. It is practicable to use a barometer 30 inches high, and further, mercury will not freeze until a temperature of $-40°$ is reached. Hence the student of the atmosphere commonly uses the mercurial barometer. Make a sketch of the barometer to be used in your observations. Letter each essential part and explain its meaning.

Reading the Barometer.—The instrument should be hung with the tube vertical. In place of the basin shown in Fig. 173, it has a smaller cup called a *cistern.* The cistern has a buckskin bottom, and by means of a screw pressing against the buckskin the capacity of the cistern can be changed. When a change of air-pressure causes the column in the tube to grow longer or shorter, the surface of the mercury in the cistern is lowered or raised. The first procedure in making an observation with the barometer is to raise or lower the mercury in the cistern by turning the screw until its surface is just at the zero of the scale. Otherwise the height of the column of mercury which balances the air at that time would be read erroneously. Note now the height of the top of the column by means of the scale, in inches, tenths, and hundredths. Record the pressure (as thus given in inches of height of the mercury column) for each day for a week or month or longer period.

In more exact or more advanced study corrections should be made for temperature and altitude, 32° Fahr. being taken as a standard of temperature, and sea-level as the standard for altitude. A thermometer is usually mounted close to the barometer, and its reading gives the temperature of the barometer. The amount of the "correction" is taken from a table (see Ward, Practical Lessons in Elementary Meteorology, pp. 166–167). If the table is not at hand, the correction may be computed approximately by the formula:

The correction (in inches) $= \frac{h(t\text{-}32)}{10{,}000},$ in which h is the barometric pressure in inches and t the temperature of the barometer in degrees. The tables for reduction to sea-level are found in Ward, pp. 168–170. Suppose your place of observation is 1,200 feet above the sea, and the temperature at the time is 80°. We find that 1.217 (inches) must be added to the reading of the barometer. This means that 1.217 inches of mercury weigh the same as the column of air reaching 1,200 feet above sea would weigh at the given temperature. Such a correction is always made in the construction of weather maps.

219. **The Mapping of Pressure—Isobars.**—Turn to Fig. 174. The isotherms are the same as those in Fig. 169, but we will not now consider these. How do the lines for equal pressure (isobars) differ in the drawing from the isotherms? How many isobars marking belts of 30 inches? Describe their position. Taking the easterly isobar of 30 inches, do the pressures decline or rise east of this? What is the lowest pressure recorded in that direction? How is the center of this area marked? Take the other isobar of 30 inches and look to the northward. How do the pressures change? What is the highest pressure recorded? Where is the area, and how is its center marked? In the same way describe the pressure conditions in the Texas region. How many centers of high and low pressure are shown on the map? Are the isobars crowded in any part of the map? What do you say here of the pressure slope or *gradient?* See pages 257, 258. The distance from Buffalo to Halifax about equals that between Parkersburg, W. Va., and Marquette, Mich. Compare the gradients in the two cases.

Compare Fig. 175, noting the dates. How has the northwest "high" changed in position from January 7th? Has the pressure changed? What change in the "low" at the northeast? What is the most conspicuous change in the distribution of pressure? Where is its center? Where is now the steepest gradient shown? Is this the same "low" that had its center on the Rio Grande on January 7th? Compare Fig. 176, noting the date. Record again the changes in the Dakota high area. What new center of high pressure is shown? What changes have come to the great low-pressure area of January 8th? Where is its center? Compare the gradients with those of January 8th. In what general direction have the "highs" and "lows" moved during these three days?

219*a*. **Making a Pressure Map.**—Take a fresh copy of the blank weather map and find in the second column of the table of data the pressures for the different stations. They are given to hundredths of an inch. Record them on the map in neat figures, as in the case of temperature.

What is the highest pressure shown? At what station? How far west do you go to find another record of 30 inches or more? What is the pressure at Atlantic City? At Washington, Pittsburg, Cleveland, Detroit, Alpena, Sault Ste. Marie, Escanaba, and Marquette? Compare the pressures at Mobile, Vicksburg, Cairo, St. Louis, Davenport, Chicago, Milwaukee, and Escanaba. Compare Bismarck, Moorhead, and Duluth. Has any considerable area pressures of 30 inches or more? What States or parts of States does it include? Does the area of such pressure approach the Pacific coast at any point?

Having thus become familiar with the general distribution of pressure, you are ready to draw the lines of equal pressure, or isobars. The method is the same as with the isotherms. Here the lines, however, are drawn for each ten-hundredths of an inch (of pressure). What is the center of low pressures? The lowest being 28.88 (at Marquette), draw the line from 28.90. As no stations have this record in this region, interpolate as before, determining the proportional distances toward Escanaba, Duluth, Port Arthur, and Sault Ste. Marie. For 29.00 you have one of these stations, and will carry the line outside of Green Bay and inside of Duluth. Continue this process until the low is sketched in, and then pass to the high area in the West. Compare the high area with the low in size, geographic position, and in its gradients. How many minor centers of high pressure do you find? Would any of these perhaps be extensive if your map gave the pressures for lands and seas beyond our borders? What is the order of succession of highs and lows as you pass from the Pacific to the Atlantic? The distribution of pressure may be brought out by the use of colors, using a separate color or shade for all areas of over 30 inches; for those between 30 and 29.50; between 29.50 and 29, and below 29.

If time permits, the temperature and pressure maps may now be combined, using a third blank weather map for the purpose and omitting colors.

220. **Winds—Local Observation.**—In the section on observing the weather without instruments brief reference was made to winds. The study should now be carried further, both without instruments, and with the wind-vane and anemometer (Fig. 177), if there is opportunity for more careful work. Note first the direction, discriminating the four cardinal and the four intermediate points, the direction being that *from which* the wind comes. For strength use the following scale:

0. *Calm.*
1. *Light;* just moving the leaves of trees.
2. *Moderate;* moving branches.
3. *Brisk;* swaying branches; blowing up dust.
4. *High;* blowing up twigs from the ground; swaying whole trees.
5. *Gale;* breaking small branches; loosening bricks in chimneys.
6. *Hurricane* or *Tornado;* destroying everything in its path.

Is the wind blowing in the same direction at all heights? Compare the motion of the lower and upper clouds to see whether it is the same. The best observations of direction are on clouds near the zenith, for they reveal the general movement of the atmosphere, while the inequalities of the earth's surface may to a considerable degree control the direction of the wind at lower levels. Observe whether the wind is persistent in one direction, or is gusty and changeable. Also note any correspondences of temperature with wind direction.

220*a*. **Winds and Atmospheric Pressures.** — At all weather stations the direction and strength of the wind are recorded as data for the weather maps. Thus the movements of the atmosphere in a region, or those of the entire country, and finally of the entire globe, can be compared. Turn to Fig. 174 and observe the directions of the wind (as shown by the arrows) in the New England and Nova Scotia region. Does the wind on the whole move toward the "low" or away from it? How is it with the area of high pressures in the Northwest? Make a similar study of the high and low areas shown in Figs. 175 and 176. Are you prepared to state a general law, as suggested by these cases? Review the illustration of the fire in the cold room and explain why the wind blows toward the "lows." This illustration does not explain the origin of the low areas, but these being given, it aids us to see why the winds blow as they do in relation to them.

You are now ready to plot on the map the data for the winds given in the table from which we have taken the temperature and pressure. Use the map on which you have drawn the isobars. Do not write the figures or letters indicating strength and direction, but use instead arrows pointed with the wind. Use a short arrow for winds having a velocity of less than 10 miles, and a longer one for those whose velocity is 10 miles or above.

Study first the direction. What is the prevailing direction in the Dakotas, Nebraska, and Kansas? In Iowa and Missouri? In Kentucky, Indiana, and Ohio? In Pennsylvania and New York? How are the directions in the areas above mentioned related to the "low" in the Superior region? If they do not point toward the "low," do they agree in pointing to the right or left of it? What kind of a movement would therefore belong to the atmosphere in the Upper Mississippi Basin and in the Lake region? Do the winds of the Gulf region belong to this system of movements? What general relation can you see between the winds and the isobars east of the Rocky Mountains? Placing yourself in northeastern Nevada, what can you say of the wind directions of the Western United States as studied from that point? Is there any exception in that region? Can you suggest a probable reason for this case? Do you find the winds related to the highs and lows in the same way as seen on the maps in the text-book?

Are the greatest velocities in the East or West? Where are the greatest differences in pressure? Is there any general correspondence between the two? Compare the distance between Boise and Portland (Ore.) with the distance between Bismarck and Duluth. (The latter is a little the greater.) What is the difference in pressure between Boise and Portland as compared with the difference between Bismarck and Duluth? Consult again the table and record the velocities at the Western pair of stations, and put with them the velocities at Bismarck and Duluth. Compare velocity at Baker City with those of Moorhead and Huron. Make similar comparisons between the Lake and Gulf regions, noting the rate of change of pressure. Do the larger velocities correspond with the larger gradients (steep-pressure slopes)?

222. **Storms of the Westerly Winds, or Cyclones.**— Refer again to Fig. 174 and mark the center of the low area on January 7, 1886, in the western Gulf region. Is it a well-developed "low"? Note its changes of position and character on January 8th and 9th. Has it followed approximately any of the paths shown in Fig. 179? What do you observe concerning the gradients and the winds when the "low" was in the western Gulf region? What changes in these respects when it reached Alabama? When it reached the Atlantic coast? Did any center of high or low pressure disappear from the map from Janu-

ary 7th to January 9th? Did any such area appear for the first time? How did the movement of the northwestern "high" compare in rate of movement with the southern "low"? Why should rain or snow be often associated with areas of low pressure? Describe the high and low areas of Fig. 178. Give the distribution of rainfall. Have we in this case any means of telling what precise path these centers were taking? Do we know from experience in what general direction they were moving? If you are in the track of a "low," what would commonly be the direction of the wind before it reached you? After it passed? What conditions of wind would you expect while it was passing over you? What change of temperature is usually connected with such a passage? What would be the successive changes in the direction of the wind if a "low" were passing eastward at some distance to the north of you? When the wind changes thus, it is said to *veer*. Describe the changes if the "low" or *cyclone* is passing south of you. In this case the winds are said to *back*. Compare these movements to those of the hands of a clock. Do they give a clue to the path of the cyclone?

225. **Shore and Gorge Breezes.**—If the school is close to the ocean or to a large lake, observe the "land and sea" breezes. This will be possible only in warm, clear weather and when the general condition of the air is quiet. Watch for a reversal of the wind in the forenoon, and again a little before or after sunset. Record the hours of observation and the strength of the breeze.

If the school is near a steep mountain front, or in a mountain gorge, a similar daily reversal of the wind can be observed. In sunshine the air against the slopes, expanded by warming, tends upward and produces a gentle current up the slopes. At night the cool, heavy air against the ground flows down the slopes and accumulates in the gorges, through which it often pours with much force.

227. **Thunder-storms.**—Observe and record the history of a storm from the moment of its approach. Describe the clouds and their changes of form, color, or motion. What are the direction and strength of the wind? How long did it rain? What was the size of the drops? How much rain fell? Did the barometer show change of pressure, and how much? Record any changes of temperature. Describe the lightning and the peals of thunder. What time passed between the flash and the peal? How long did the storm last? Write out in simple narrative form the story of the storm. Compare it with other thunder-storms which you remember.

232. **Weather Maps.**—Review sections 232 and 234. You have already given attention to such maps in exercises 216, 219, 220, and 222, to the temperatures, pressures, and winds already plotted from our table of data. You can now add the rainfall, placing the figures by the various stations as before. Areas of rainfall may be shown approximately by shading, where several rainy stations appear adjacent to each other. You will now have completed your weather map for this one day. Give a concise description of it, referring to the several weather elements as related to each other.

If sufficient time is available, construct weather maps for six successive days from data given by Ward in folder opposite page 90, and discuss the weather changes of the whole period. Or select from the current Government weather maps a series for any week, and study each day's weather, and record the changes from day to day.

Apply the principles learned throughout the study to the weather elements daily observed in your own locality, and seek to forecast rainfall, direction of winds, and changes of temperature.

CHAPTER XII

THE EARTH'S MAGNETISM

237. **Declination of the Needle.**—If a theodolite or transit is available, make an observation on Polaris to determine the declination of the needle. Adjust the instrument and set it up in a convenient place before dark. Take special care in leveling it. A faint light held near the object-glass of the telescope will illuminate the cross-wires after dark. The star Polaris is not at the north pole of the sky, but a little to one side of it in the direction opposite to that of the star Alioth. In the constellation popularly known as the Dipper the star of the handle, which is nearest the bowl of the dipper, is Alioth. Point the telescope at Polaris when Alioth is either directly below it or else has the same height at right or left. Then read the direction indicated by the compass needle attached to the instrument. If Alioth is below Polaris, this direction is the declination of the needle. If Alioth is at the right, the direction is too far to the west; if at the left, too far to the east. The correction to be applied varies in amount with latitude of the place, being $1\frac{1}{2}°$ in latitude 37° and $1\frac{3}{4}°$ in latitude 47°. From July to September the observation can be most conveniently made with Alioth at the left; from October to December, with Alioth below; and from January to March, with Alioth at the right.

Isogonic Map for the United States in 1902 (Fig. 189). Describe the course of the line of no declination.

What States are wholly or partly in the area of west declination? What is the highest west declination shown? The highest east declination? How does the declination in New England compare with that in New Mexico? Compare eastern and western New York. What is the range of declination in California? In what part of the United States does the declination change the most in going from place to place? What is the declination for your own locality? Get the magnetic north and determine the position of the true north.

238. **Magnet and Compass.**—To perform the experiment mentioned in the text it is best to use a straight magnet instead of one with the ordinary horseshoe form. Such magnets are made of tempered steel. A bar of soft iron, if placed parallel to the earth's axis, derives magnetism from the earth, and, if large enough, will control a compass needle in a similar way.

239. **Compass of Iron Ore.**—Bring a pocket compass near to different parts of a block of magnetic iron ore and find its poles. Its north pole, or poles, will be found to attract the south end of the compass needle, and *vice versa.* Place the block in a pan and float it on water. It will come to rest in a particular position controlled by the earth's magnetism.

CHAPTER XIII

THE OCEAN

240. **Study of Ocean Basins.**—The details of this exercise will depend on the maps or reliefs which are at hand, and must be determined by the teacher. If the Atlantic, vol. ii, by Thomson, is available, use the folding map of the Atlantic basin, opposite the title-page. Describe the areas which are more than 3,000 fathoms in depth. What are the forms and extent of the areas under 2,000 fathoms deep? Compare the continental shelf of North America with that of Europe. What islands lie within it in each case? What land-locked seas lie behind these islands? How are the deeps related to the central portions of the North and South Atlantic? Describe the central belt as to depth, width, and changes of direction. Compare the depths of the Mediterranean and Caribbean. Compare the continental shelves of West Africa and eastern South America. Make a profile of the bottom of the North Atlantic from Porto Rico, along the track of the Challenger, past the Canaries to the African coast. Use horizontal scale of the map and vertical scale, $\frac{1}{16}$ inch = 500 fathoms. What is the latitude of this course? Consider it as 20°, though the average is a little higher. How long is a degree of longitude here? See table, page 4. What distance does your profile represent? What is the amount of vertical exaggeration? Mark off on a circle an arc corresponding to this fraction of the earth's circumference. Reproduce your profile with diminished vertical exaggeration on this curve. Make a profile as

above from Cape St. Roque to the coast of Sierra Leone. From Rio Janeiro directly eastward to the coast of Africa. Compare the three profiles. If this map is not at hand, the questions and directions will serve in a general way as a guide to the study of other maps.

243. **Islands in the Ocean.**—Review the section. Use Fig. 195. Make profile from right to left, embracing the larger and the smaller mountains, the barrier reef, and the ocean bottoms. As we have no soundings, the work must be of a general nature. Make the bottom shallow in the lagoons and deeper outside the reefs. Make profile in a similar way, using the atoll of Fig. 196.

If you have Dana's Corals and Coral Islands, turn to the map, plate xi, opposite page 204. Or take United States Coast Survey Chart No. 1,002, showing Straits of Florida and approaches. From either of these study the Bahama Islands in reference to the adjoining banks, to the deep-water channels on the south and west, and to the open sea on the east. How much land would there be in the islands if the sea were to subside or the lands rise a distance of 50 feet? Make a written description of the various features and of such possible changes.

244. **Properties of Sea-water.**—Dissolve salt in water until the water is saturated. The resulting brine corresponds approximately to that of Great Salt Lake. Mix 1 pint of this brine with 6 pints of fresh water. The mixture corresponds in saltness to sea-water. Select some object (a block of heavy wood, for example) which will barely float in fresh water. Remove it successively to the sea-water mixture and the strong brine and observe how much higher it floats on the heavier liquids. Repeat the experiment with a block of ice representing an iceberg. On a very cold night expose equal amounts of three liquids in vessels of the same size; in the morning observe the differences in the formation of ice.

249–250. **Tidal Observations.**—If the sea or a tidal estuary is near, establish a tide-gage and make regular observations. Choose a spot not reached by large waves, and having graduated a board in feet and inches, and numbered the foot-marks, fasten it vertically on a wharf or pile with the zero below lowest water. Observe the height of the water hourly, or as often as convenient, for several days, and plot the height as a curve. Then observe the times of high water and low water daily and compare them with the time of the rising of the moon, either observing the moon rise or taking it from an almanac. Having learned how much on an average the tides precede or follow the rising of the moon, predict the times of high and low tide for future days, and then compare observation and prediction. Do the same for spring and neap tides, comparing them with full and new moons.

251. **Ocean Currents.**—Fig. 201. Take a sheet of paper of twice the dimensions of this page, and lay out the latitude and longitude with scale double that used in the figure. Draw the outlines of the continents. Draw the lines and arrows for the equatorial current (using red for the ocean currents), the Guinea Current; the Gulf Stream; the Labrador Current; the South Atlantic Currents. Record the dates and general course of the bark Telemach; the same for the schooner W. L. White. On the Gulf Stream, read, if at hand, Thomson, Depths of the Sea, chapter viii, p. 356, etc. Also Three Cruises of the Blake, vol. i, chapter xi, p. 241, with map, Fig. 174. Compare Franklin's idea of the Gulf Stream, Fig. 173. Read also in Pillsbury's Gulf Stream, published by the United States Coast Survey. Give a brief written report from such readings.

Take any pilot chart of the North Atlantic or other chart from the United States Hydrographic Office which shows drift of bottles in the Atlantic Ocean. The following questions are based on chart of August, 1898, title Recent Accessions to our Knowledge of Ocean Currents Obtained Through Floating Bottles. Read carefully the account under this title at the right side of the charts and follow the map in the reading. Observe that each bottle track is numbered, and that the numbers on the map correspond to numbers in the tabular description in the central part of the sheet. What symbol indicates the point where the bottle was set adrift? What shows where it was recovered? What numbers indicate the eastward-moving equatorial current? On what coast were the bottles recovered? When was bottle No. 76 thrown over? When recovered? Distance traveled? Average drift per day? Is this rate of drift high or low? What are the highest and lowest rates in the table? Would a bottle be certainly recovered as soon as it reached land? If not, how would this affect the estimate of rate of drift? Trace and describe the course of No. 3; of No. 1; of No. 65. Compare No. 3 with No. 23. Compare No. 1 with No. 81. What do these comparisons show? What is the general course of bottles set adrift west of North Africa? The same for those set adrift north of 40° north latitude? Trace the course of No. 7 and No. 37, drawing conclusions.

If the above chart is not available, take any similar chart of the Hydrographic Office and study the drift of bottles in a similar manner.

Study the United States Hydrographic Chart, The Derelict Schooner W. L. White (supplement to the Pilot Chart of the North Atlantic Ocean for February, 1889). Describe the course of the derelict, giving dates with latitude and longitude of point where sighted. Observe that this appears in Fig. 201, but without dates, owing to small scale. Read the history of the wreck as printed on the chart. Record the chief facts in your note-book. How long a time did the lone voyage require? When and where did she come ashore? How many reports reached the Hydrographic Office? Where was she June 21st? What was the report for that date? What is the value of such records? What other derelicts are charted here?

252. **Ice in the Sea.**—Chart—Ice in the North Atlantic—Season of 1889–90, supplement to the Pilot Chart of the North Atlantic Ocean, January, 1891, United States Hydrographic Office.

What is the symbol for icebergs? For field or floe ice? For how many months have we a record? Compare the area here shown with a general map of the North Atlantic. Plot this area on such a map. In which months is this region most nearly free from ice? When are the icebergs most numerous? Compare their distribution in January, 1890, with that of February, 1890. What does the printed statement at the top of the chart say of the apparent absence of ice from the Grand Banks? When does the field ice begin to be abundant? How long does it continue plentiful? Describe its distribution about Newfoundland in June, July, and August. Why is the ice most abundant in the spring and summer months? Is this an average year for ice in the North Atlantic? See notes at the top of the chart. What feature of Fig. 201 is suggested by the distribution of the ice?

253. **Exploration of the Sea.**—This exercise will not be outlined here, but the student may be assigned readings for written reports, the material depending upon the volumes available, such as Nansen's Farthest North, Greely's Three Years of Arctic Service, these especially in connection with sea ice as above.

For more southern seas and the sea bottoms, as well as for methods of sounding, dredging, etc., see Agassiz's Three Cruises of the Blake, Thomson's Atlantic, Thomson's Depths of the Sea, and Captain C. D. Sigsbee's Deep Sea Sounding and Dredging.

256. **Navigation.**—Pilot charts of the North Atlantic Ocean, August, 1889, and February, 1900, United States Hydrographic Office. If at any time these are not available, two others representing summer and winter seasons should be used, with questions similar to those that follow. Use the August chart. What topics or kinds of explanation are given in the printed matter on the right and left sides of the charts? What is the hurricane signal? The N. W. wind signal? For what regions were the new charts published in July, 1899? What branch hydrographic office is nearest your home? How are storm tracks shown and how is varying intensity indicated? How is the ice later seen distinguished from that earlier seen? How are regions of frequent fog designated?

Describe with reference to latitude, longitude, and adjacent lands, the northern steamship routes from New York and Boston to Europe. What limits of dates are advised in the case of vessels moving eastward and westward? What advice is given to sailing vessels? Describe the southern route, and indicate when it is to be used. Where is the southern route for west-bound steamers from Gibraltar? When is the northern route from Gibraltar to New York to be used? Describe the track and intensity of storm No. 1. Describe the region of equatorial rain as shown on this chart. What is indicated by the fine dotted lines? By the circles with figures and arrows? By the small arrows?

Using also the chart for February, 1900, compare the various features with those of the previous summer. Compare the direction and intensity of the winds along the track of the steamships from New York and Boston. The same along the southern route for west-bound steamers from Gibraltar. Note the relative amounts of sea ice in the two seasons; also the fog predictions for the two periods. Compare the areas of equatorial rain and the limits of the trade-winds.

257. **Submarine Cables.**—Map—Submarine Cables of the World, etc. National Geographic Society.

How many lines are indicated as crossing the North Atlantic? What chief points are touched on the American side? How many cables touch Ireland? Land's End? France? What are the principal termini of lines in the Mediterranean? What important lines run from Lisbon? What connection have the Bermuda Islands? What important lines south of Asia? Describe the location of lines for Australia and New Zealand. What important lines do not appear on this map of the Pacific region? Observe the land connections. What important line in North America is not here shown? Any important omission for Africa? Judging by telegraph-lines, how fully is the world brought under the dominion of civilization? The student is reminded that only the "principal" connecting land lines are here shown.

CHAPTER XIV

THE MEETING OF THE LAND AND SEA

259. **Shore-line of Maine.**—Write a general description from any atlas map. For detailed study use the Boothbay and Bath Sheets, taking first the former. What are the township or other names on the chief headlands? What are the intervening bays? What river enters the sea here? How far inland do the bays penetrate? Does this map show their full extent? What is the height of the headlands? How would the proportions of land and water in this area be changed if the region were to sink 100 feet? What islands would disappear? What headlands would become islands? If you have Coast Survey Sheet 105 or Harbor Sheet 314, study the soundings about the mouth of the Penobscot. How much would the land be increased if the region rose 100 feet (estimate)? What islands would become a part of the mainland?

Make a similar study of the Bath Sheet. Taking both, what culture features do you find that would not appear on an inland map? Make a list of the lights. How do the roads differ from those about Fargo (in Minnesota and North Dakota)? Why? How does topography affect the making of railways in this region? How are the settlements located, and why? Where are the country and township lines usually found? Discuss the origin of such deep, narrow bays, taking account of rivers, glaciers, and the sea.

260. **Shore-line of Massachusetts.**—Several exercises are suggested in the Teacher's Guide, as the entire shore-line of the State abounds in interest. The Coast Survey Chart No. 7, Cape Ann to Block Island, should be examined, and a brief general description of the coast written. For detailed work take Boston and Boston Bay Sheets. Study the headlands and beaches. Describe the relations of Marblehead Neck to the mainland. The headland is rocky and but thinly covered with drift. The "neck" proper is a sandy beach. Has the headland always been joined to the mainland? Describe Nahant. Show the effect of waves and currents on the shore-line from Lynn by Grow's Cliff to Point Shirley. What would you infer from the topography as to the former conditions in the vicinity of Hull and Nantasket? What is the origin of the marshes west of Revere Beach? Make a list of the lights shown on this sheet. What width of open water leads from Boston Harbor to Boston Bay? Was there formerly more, or less? What rivers enter Boston Bay and Boston Harbor? What is their character? (The student is reminded that the tidal marshes were formerly more extensive, since much of Boston stands on "made" land.) If possible, study Coast Survey Chart 246 or 337, noting the soundings. Describe the changes that would follow an uplift of the land amounting to 100 feet.

261. **New York and New Jersey.**—Special sheet, New York and vicinity, and Coast Survey Chart No. 8. Read the section and examine Fig. 208. This cut is made from a photograph of a relief model. The model is based on the special sheet now to be used for study.

What are the dimensions of the area shown by the map? Describe the Coney Island and Rockaway Beaches in their relation to Long Island. In what direction have their sands been moved? Explain the origin of the marshes about Jamaica Bay. What is the origin of

Sandy Hook? Why are Navesink and Shrewsbury Rivers so wide at their lower courses? What other rivers of the map show the same feature? What is the ridge which separates the Hudson River and the upper bay from the Hackensack Valley? What is the length and approximate area of the marshes between Hackensack and Perth Amboy? Why are the shore-lines along the East River and Long Island Sound so ragged? Make a list of the lights. How many life-saving stations do you find?

Study Chart No. 8, United States Coast Survey. How does Great South Bay correspond with Jamaica Bay? Does the form of Fire Island Beach corroborate your explanation of Rockaway and Coney Island Beaches? What is the most eastern land shown on this map? The most southern? How far is an offshore beach found to the southward of Sandy Hook? Does it appear in the vicinity of Long Branch and Asbury Park? How many inlets between Seaside Park and Cape May? Compare the inner and outer shore-lines southwestward from Montauk Point. How do the soundings in the Hudson range from Jersey City to Yonkers? What do the figures indicate when on the dotted surfaces? What are the limits of depth in the upper bay? Find a track across the lower bay for a vessel drawing 30 feet of water. Describe the course of the 10-fathom line. Study the position of the 20-, 30-, 40-, and 50-fathom curves. What course would the waters of the Hudson take if the land were uplifted 300 feet? Could the sea make such a channel as you now find on its bottom? Could the sea cut out the basin of the upper bay or the lower channel of the Hudson? How could these features be made? How far from the south shore of Long Island is the water less than 100 feet deep? How far from the New Jersey shore-line? How far is the 100-fathom curve from New York? How far is the 1,000-fathom curve? What does this difference mean? How many lights are shown? How many light vessels? What is the range of distance for which they are visible?

262. **South Atlantic and Gulf Coasts.**—Materials will depend on the locality preferred for study, and the exercises should be carried out in a way similar to those given above.

For the Chesapeake Bay region use a group of the United States topographic sheets and Coast Survey Chart 79 or 9. The former gives Chesapeake Bay, the latter the coast from Cape May to Cape Henry. Chart 131 gives Chesapeake Bay entrance, Hampton Roads, etc. For the North Carolina coast, Charts 10 and 11. For the greater part of South Carolina and Georgia, including Charleston and Savannah, Chart 12. No. 154 gives Charleston and vicinity, and 431 Charleston Harbor. No. 155 gives Savannah Harbor and the adjacent coast. For Straits of Florida, Key West, etc., see Chart 15. No. 18 gives Mobile, 19 the Mississippi Delta, and 21 Galveston. No. 188 gives Mobile Bay, No. 194 the mouths of the Mississippi River, and No. 204 Galveston Bay.

263*a*. **Pacific Shore-line.**—San Francisco and vicinity. Coast Survey Chart No. 5,500, Point Pinos to Bodega Head. What distance on this map represents a statute mile? When were the latest corrections of this chart made? When were the first surveys made for this chart? How are soundings for the shallow waters expressed? What is the plane of reference from which these soundings are determined? What are some of the characters of the bottom shown by the map? What sign indicates a sunk rock? A life-saving station? What is the declination of the needle, as shown by the chart? What are the soundings in the Golden Gate? How wide is this passage? What are the soundings outside of the Golden Gate along the "main ship channel"? What is the average distance from the shore of the 30-fathom curve? Of the 100-fathom curve? How can you account for the shallowing outside of the Golden Gate? Where are the deeper waters of San Francisco Bay? What rivers discharge through the northern arm of the bay? Consult Fig. 129 for review of California topography, also Fig. 216.

263*b*. **Coast Survey Chart No. 3,089.**—Inland passages, Olympia, Wash., to Mount St. Elias, Alaska. What is the length of this shore-line in statute miles by direct course? Estimate the length, including all shores of islands, bays, and headlands. What declinations of the needle are shown on this map? What are the soundings at Seattle? At Tacoma? At Victoria? In Georgia Strait and Discovery Passage? What symbol indicates steamer routes? Trace the inside passage from Georgia Strait to Glacier Bay, Alaska? Give soundings and width of passage by Princess Royal Island through Grenville Channel and Frederick Sound. What are the deepest soundings in Lynn Canal? What are the soundings off Point Manly in Yakutat Bay? What amount of elevation of the land would cause a retreat of the waters from most of these fiords?

271–272. **Coast-lines of Rising and Sinking Lands.**—The Harvard Geographical Models or Figs. 10, 15, and 16 of this manual. Take Model 1 or Fig. 15. How is *relief* shown in this model? Has much or little wear taken place over these lands? Do the mountains show the same form to the right as to the left? How do opposite slopes compare with each other on the right? On the left? What structure have the rocks on the left? (See section 4.) Are they as originally deposited? Taking the top as north, what causes the ribbed appearance on the northeast slope of these ridges? What share of the whole land area is drained by one master stream? Has this river a flood-plain? What does this flood-plain become near the sea? What work has the sea accomplished on the edge of the land? Along what share of this coast-line is

FIG. 15.—Harvard Geographical Model, No 1.

such an effect produced? If you have the descriptive pamphlet which accompanies these models (Proceedings of the Boston Society of Natural History, vol. xxviii, No. 4, pp. 85–110), read pages 89–101.

FIG 16.—Harvard Geographical Model, No. 3.

Model 2 or Fig. 10. Review exercise 150*a*. Compare the lands and rock types in Fig. 10 with those of Fig. 15. What change has taken place in the relations between land

and sea? What has happened to the delta of Fig. 15? Where are the shore-cliffs of Fig. 15? What changes have been wrought upon the uncovered sea bottom? Where is the present delta? What features here correspond to features of the New Jersey shore-line? What feature of the lands shown here does not appear in Fig. 15? In the pamphlet read pages 94–101.

Model 3 or Fig. 16. Compare the mountains on the east and west. Where is the delta? Where is the coastal plain? Are any remnants of the shore-cliff in sight? What is the origin of the islands? Which condition would be most favorable for population, that of Fig. 10 or Fig. 16? What industries would be prominent in the former case? In the latter? Name shore-lines in any part of the world like that of Fig. 16? What name is properly applied to such a coast? In the pamphlet read pages 104–107.

Imagine this region to remain with this relation to the sea for a long time. Name the changes that would take place. Imagine an uplift to follow to the position of Fig. 10. Would the resulting shore-line and adjacent lands be the same as in Fig. 10? How would they differ? Make a list of all the forces that are involved in the changes of land form considered in this exercise.

CHAPTERS XV AND XVI

THE EARTH AND ORGANISMS

278–279. **Forests and Forestry.**—In this and subsequent topics the student will look to the teacher for references to suitable readings and for guidance in study. The accompanying maps, Fig. 17 of this manual, give the distribution of forest reserves and national parks to June, 1900. How many forest reservations appear on the map? How many national parks? Where are the latter and what are their names? In John Muir's volume, Our National Parks, read the author's account of the Yosemite Park, and write a short paper, giving the main facts. How many forest reservations in California? Where are they located? What are the reasons for thus protecting forest lands? Where are the reserves of Oregon and Washington? Can you trace any relation between these forests and the rainfall? See map, Fig. 160. How many reserves in Colorado, with their names? Account for the presence of forests and a reserve in northern Utah? To what mountain range or group of ranges do the reserves of Wyoming, Montana, and Idaho belong? Why is there no reserve in Nevada?

Practical Forestry, by John Gifford. If you have at hand, turn to pages 230, 231, and answer the following questions: Which is the largest reserve in Arizona? What is its area? When was it set apart as a reserve? What is the total area of all the reserves in Arizona? What is the largest reserve in the entire West? In what State? When was the reserve established? What reserve in South

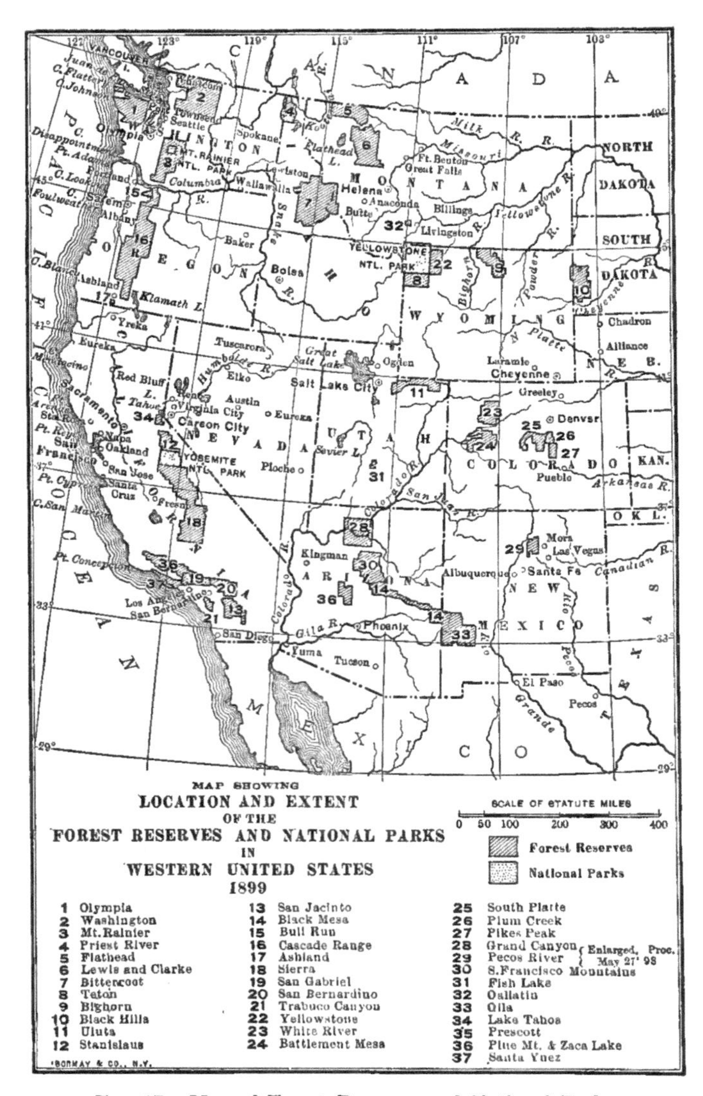

FIG. 17.—Map of Forest Reserves and National Parks.

Dakota? What connection between the forest and other geographic conditions?

Distribution of Plant Life in the Home Neighborhood.—In what ways has man changed the distribution of the plants in your neighborhood? Has he cut away the forests? What proportion of the ground is yet covered with forest? What weeds make it difficult to keep your lawn in good condition? What are the troublesome weeds of the garden and the fields? Have any been more difficult to cope with in the last few years? In the field do you think it more effective to remove the weeds by cultivation or to suppress them by heavy cropping. Have you had experience of setting nursery shrubs or fruits that winter killed? How did the climate of the nursery differ from your own? Name cases in which plant life was overcome by other plants. Name cases in which the greatest foe was browsing animals or insect pests. Read in this connection section 297, and Chapter IX in Coulter's Plant Relations. Does the term "struggle" apply quite as well to plants as to animals? Study the natural vegetation of your neighborhood. Visit the nearest swamp. What trees and shrubs or herbaceous plants do you recognize? Are there any which you do not find on the dry lands about? If the swamp borders a pond, lake, or river, see if you can find a belted arrangement of plants similar to that seen in Fig. 222. What shrubs do you find along the roads and fences? What trees make up the greater number in the nearest wood-lot? What trees are only occasional or rare? Is there thick undergrowth in the forest? If not, is the condition due to pasturing or to suppression by a heavy stand of mature trees? Do you know farmers that take intelligent care of their woodlands? Are they careful to cut the ripe timber? Do they avoid careless destruction of growing trees? Do they thin the timber, preserving the more symmetrical trees?

What kinds of evergreen trees are found in your neighborhood? Can you distinguish any special features of soil or moisture where they flourish? Have you noted any differences between trees of the same kind, comparing those which are in the forest with those which grow in open fields?

285–287. **Distribution of Animals in North America.**—Figs. 226 and 231. Take the former. Describe the distribution of the musk-ox. What color is used as a symbol? What is the most southern latitude reached? The most northern? What parts of the United States have the moose? What is its eastern limit? How does the belt of occupation run with reference to the Great Lakes? With reference to Hudson Bay? In what great river basins does this animal appear? Where does it reach the Pacific Ocean? Does it touch the Arctic Sea? What countries share in the distribution of the antelope (in 1900)? In what country is the largest area? What physiographic regions of the United States embrace most of the antelope area? What is the eastern limit in longitude?

In Fig. 231 to what extent do the animals named confine themselves within national boundaries? Note exceptions in each case. How does the range of the woodland caribou compare with that of the moose (Fig. 226)? Compare the areas of antelope and the deer. Compare the distribution of the barren-ground caribou and that of the musk-ox.

296. **Tendency to Spread.**—Map, Fig. 236. How many States received the sparrow by the hand of man between 1852 and 1870? Into how many additional States had it spread down to the year 1886? Into what States did man carry the bird from 1870 to 1886? What new areas had been covered by 1898? What areas of distribution lie west of Kansas? Does the bird seem to be dependent on any special set of geographic conditions?

APPENDIX

UNITED STATES CONTOURED MAPS

THE following list embraces all that are named in the Manual. Sheets strictly essential to the respective exercises are given first. Sheets useful but less important follow, in parentheses. Special sheets are bracketed.

Arizona.—Kaibab, Echo Cliffs.
California.—[Mt. Shasta.]
Colorado.—Leadville.
Illinois.—Savanna, Clinton, Dunlap, (La Salle).
Kansas.—Kinsley, (Meade, Dodge, Coldwater).
Louisiana.—Donaldsonville, East Delta, (West Delta, Thibodeaux, Houma).
Massachusetts.—Plymouth, Belchertown, Northampton, Springfield, Boston, Boston Bay.
Maryland.—Wicomico.
Minnesota.—Fargo, Minneapolis, St. Paul.
Missouri.—Kansas City, Marshall, St. Louis East.
Nebraska.—(Wood River, Lexington, Kearney.)
Nevada.—Disaster, Granite Range.
New York.—Elmira, Oriskany, Wilson, Tonawanda, Oswego, Mt. Marcy, Catskill, Kaaterskill, Tarrytown, (Watkins, Clyde, Geneva, Baldwinsville, Auburn, Paradox Lake, Oneida), [Niagara Falls, New York and Vicinity].
Ohio.—(Cincinnati; East Sheet, West Sheet.)
Pennsylvania.—Harrisburg, Pine Grove.
Rhode Island.—Charleston, Stonington.
Tennessee.—Chattanooga, (Sewanee, Cleveland, Ringgold, McMinnville).

Utah.—Tooele Valley.
Virginia.—Monterey, (Piney Point).
West Virginia.—Charleston, (Huntington, Wheeling, Kanawha Falls, Nicholas, Oceana, Raleigh, Hinton).
Wisconsin.—Eagle, Sun Prairie, (Waterloo, Watertown, Evansville, Stoughton, Koshkonong, Whitewater).

Summarizing, we find of the more necessary sheets, 43; of the supplementary, 37; of the special, 3. The following is suggested for an average equipment:

43 sheets, sets of 12 each, 516 sheets, at 2 cents	$10.32
37 sheets, sets of 3 each, 111 sheets, at 2 cents..........	2.22
New York and Vicinity, special (included in wholesale orders, 10 cents each, 25 cents regular), 12 copies...	1.20
Niagara Falls and Vicinity, special (regular rate, 10 cents, wholesale, 4 cents), 12 copies.....................	.48
Shasta, special (regular, 5 cents, wholesale, 2 cents), 12 copies......	.24
Total..............................	$14.46

The number of the maps can be made greater or less, to accord with local needs. For durability, maps in regular use should be backed with common muslin. Stretch the muslin on a surface of soft wood and fasten with tacks, dampen cloth and map, apply boiled flour paste, and smooth with dry cloth, or better, with photographer's roller. Do not remove until thoroughly dry, then trim cloth to margin of the sheet.

INDEX